SHADOWS
OF AFRICA

SHADOWS

Peter Matthiessen

HARRY N. ABRAMS, INC., PUBLISHERS, NEW YORK

OF AFRICA

Mary Frank

Project Director: Robert Morton
Editor: Harriet Whelchel   Designer: Samuel N. Antupit

Library of Congress Cataloging-in-Publication Data

Matthiessen, Peter.
  Shadows of Africa/Peter Matthiessen; illustrator, Mary Frank.
  p. cm.
  ISBN 0–8109–3828–6
1. Natural history—Africa—Outdoor books. 2. Zoology—Africa.
3. Wildlife conservation—Africa. I. Frank, Mary, 1933–. II. Title.
QH194.M38 1992
508.6—dc20   92–5246

Illustrations copyright © 1992 Mary Frank. Mary Frank
is represented by Zabriskie Gallery, New York

Text copyright © 1972, 1981, 1991, 1992 Peter Matthiessen

The African journals in this book were excerpted from the
sources given below and are reprinted with the generous
cooperation of the publishers: *The Tree Where Man Was Born*
(Dutton, 1972), *Sand Rivers* (Viking, 1981), and *African Silences*
(Random House, 1991). Additional sources are "Botswana,"
in *Audubon Magazine* (January 1981), and "Tanzania," in
Audubon Magazine (May 1981). Portions of *The Tree Where Man
Was Born* and *Sand Rivers* appeared earlier in *The New Yorker*.

CONTENTS

To the courage and dedication of the rangers and wardens of the African national parks

With appreciation to my son, Pablo,
for his endless curiosity and interest in the
world of animals; and for Leo—M. F.

In fervent hope that my small grandsons,
Christopher, Joseph, and Andrew, together
with all of the world's children, may still
find remnants of the splendor of the world
when they grow up—P. M.

PART 1

S O U T H S U D A N

The tree where man was born, according to the Nuer, still stood within man's memory in the west part of the south Sudan, and I imagined a great baobab thrust up like an old root of life in those wild grasses that blow forever to the horizon, and wild man in naked silhouette against the first blue sky. And I imagined that this glimpse of the earth's morning might account for the anticipation that I felt, the sense of origins, of innocence and mystery, like a marvelous childhood faculty restored. Perhaps it is the consciousness that here in Africa, south of the Sahara, our kind was born. But there was also something else that, years ago, under the sky of the Sudan, had made me restless, the stillness in this ancient continent, the echo of so much that has died away, the imminence of so much as yet unknown. Something has happened here, is happening, will happen—whole landscapes seem alert.

We drank the black tea of the desert and, in late twilight, started off again, traveling onward until after midnight, when once again we lay down upon the ground. The night was warmer, warm enough for the mosquitoes, and it came to an end at last. During the night, the hippos bellowed from the Nile, a distant sound, the first murmurings out of the heart of Africa

The first light shone on a new land of long grass and small acacia, with occasional great solitary baobab. The feather-leaved, sweet-scented acacias or thorn trees, in their great variety, are the dominant vegetation in the dry country south of the Cape, but the tree of Africa is the baobab, with its gigantesque bulk and primitive appearance; it is thought to reach the age of 2,500 years, and may be the oldest living thing on earth. The grassland danced with antelope and birds—tropical hawks, doves, pigeons, guinea fowls and francolins, bee eaters, rollers, hornbills, and the myriad weavers, including the quelea, or Sudan dioch, which breeds and travels in dense clouds and rivals the locust as an agent of destruction. At the edge of a slough stood two hundred crested cranes and a solitary ostrich, like a warder; where trees gathered in a wood were the white faces of the vervet monkey. In the afternoon, the savanna opened out on a great plain where gazelles fled to the horizons, and naked herdsmen, spear blades gleaming, observed the passage of the truck through the rushing grass with the alert languor of egrets. All the world was blue and gold, with far islands of acacia and ceremonial half circles of human huts.

Nimule

Nimule is the only national park in the Sudan, a natural park between the mountains and a bend in the Albert Nile. To the south and west, early one morning, the mountains of Uganda brought the sky of Africa full circle. Somewhere in those mountains, down to the southeast, lived a light, small people called the Ik who recently used pebble tools of the sort made in the Old Stone Age; in the Congo's Ituri Forest, to the west, lived Pygmies who still carried fire rather than make it.

Soft hills inset with outcrops of elephant-colored boulders rose beyond a bright stretch of blue river, and elephants climbed to a sunrise ridge from a world that was still in shadow. More than a hundred moved slowly toward the sun.

On the west bank, the askaris [rangers] shook small bags of fine dust to gauge the direction of the wind. We moved inland. Very soon there arose out of a copse a herd of buffalo, with its coterie of cattle egrets rising and settling once again on twitching, dusty backs. To judge from the rapidity with which the askaris cocked their rifles, we were too close; the beasts took a few steps forward. Wet nostrils elevated to the wind, they wore an aggrieved, lowering expression. There were no handy trees to climb, and I wondered how to enter most promptly and least painfully the large thornbush close at hand. But the buffalo panicked before I did, wheeling away in the dark commotion, leaving the white birds dangling above the dust.

To the south, on a rise that overlooks the Albert Nile where it bends away into Uganda, a herd of kob antelope stepped along the hill—some sixty female kobs and calves led by a single male with sweeping horns and fine black forelegs—and the delicate oribi, bright rufous with brief straight horns, scampered away in twos and threes, tails switching. A gray duiker, more like a fat hare than an antelope, gathered its legs beneath it in low flight, and a sow warthog with five hoglets, new sun glinting on the manes and the inelegant raised tails, rushed off in a single file at the scent of man.

To the east, the entire hillside surged with elephant, nearly two hundred now, including a few tuskers of enormous size. And to the north, on a small hillock, stood four rhinoceros, one of these a calf. The askaris approached the rhinos gradually, keeping downwind—not always a simple matter, as the light wind was variable—and eventually brought us within

stoning distance of the animals. The rhinos were of the rare "white" (from the German *weit,* or wide-mouthed) species, a grazing animal that lacks the long upper lip of the black rhino, which is a browser; mud-crusted with their double horn, their ugliness was protean. The cow and calf having moved off, two males were left, and these, aware of an intrusion but unable to detect it, moved suspiciously toward each other, stopping short at the last second as if to contemplate the risks of battle, then retreating. The white rhino is gentle and rarely makes a charge; buffalo in herds are also inoffensive, and no doubt the askaris were teasing as well as pleasing us, though they kept their laughter to themselves.

Beyond the rhino, dry trees rose toward the dusty mountains, and beyond the hills hung the blue haze of Africa, and everywhere were birds—stonechats and silver birds, cordon bleus and flycatchers, shrikes, kingfishers, and sunbirds. Overhead sailed vultures and strange eagles and the brown kite of Africa and South Asia, which had followed me overland two thousand miles from Cairo, up the Nile. Here in Equatoria, in the heart of Africa, with Ethiopia to the east, Uganda and the Congo to the south, Lake Chad and the new states of what was once French Africa to the west, one sensed what this continent must have been, when the white rhinoceros was not confined to a few pockets but wandered everywhere, like the kites, from the plains of Libya south to the Cape of Good Hope. Today Libya is desert, and the wild things are disappearing. The ragged kite, with its affinity for man and carrion, will be the last to go.

In 1961 the great animal herds of the Serengeti Plains seemed on the point of disappearance. Traveling overland from Cairo, I got as far south as the Ngorongoro Crater, in the country still then known as Tanganyika. The new Sunday air charter from Nairobi permitted a brief visit to the Serengeti, where I saw the first leopard of my life, loping along among low bushes by a stream; on the homeward journey the pilot flew over the endless companies of game animals on the plain that is the greatest wildlife spectacle left in the world.

In 1969 and 1970, I spent most of the winter in the Serengeti as a guest of the Tanzania National Parks, exploring the park in my own Land Rover and accompanying the warden and the scientists of the Serengeti Research Institute on expeditions into the field.

Serengeti

One morning a great company of elephants came from the woodlands, moving eastward toward the Togoro Plain. "It's like the old Africa, this," the Serengeti Park warden Myles Turner said, coming to fetch me. "It's one of the greatest sights a man can see."

We flew northward over the Orangi River. In the wake of the elephant herds, stinkbark acacia were scattered like sticks, the haze of yellow blossoms bright in the killed trees. Through the center of the destruction, west to east, ran a great muddied thoroughfare of the sort described by the elephant hunter F. C. Selous in the nineteenth century. Here the center of the herd had passed. The plane turned eastward, coming up on the elephant armies from behind. More than four hundred animals were pushed together in one phalanx; a smaller group of one hundred and another of sixty were nearby. The four hundred moved on in one slow-stepping swaying mass, with the largest cows along the outer ranks and big bulls scattered on both sides. "Seventy and eighty pounds, some of those bulls," Myles said. (Trophy elephants are described according to the weight of a single tusk; an eighty-pound elephant would carry about twice the weight in ivory. "Saw an eighty today." "Did you!")

With their path-making and tree-splitting propensities, elephants will alter the character of the densest bush in very short order; probably they rank with man and fire as the greatest force for habitat change in Africa. In the Serengeti, the herds are destroying many of the taller trees, which are thought to have risen at the beginning of the century, in a long period without grass fires that followed plague, famine, and an absence of the Maasai. Dry-season fires, often set purposely by poachers and pastoral peoples, encourage grassland by suppressing new woody growth; when accompanied by drought, and fed by a woodland tinder of elephant-killed trees, they do lasting damage to the soil and the whole environment. Fires waste the dry grass that is used by certain animals, and the regrowth exhausts the energy in the grass roots that is needed for good growth in the rainy season. In the Serengeti in recent years, fire and elephants together have converted miles and miles of acacia wood to grassland and damaged the stands of yellow-bark acacia or fever tree along the water courses. The range of the plains game has increased, but the much less numerous woodland species such as the roan antelope and oribi become ever more difficult to see.

Ordinarily the elephant herds are scattered and nomadic, but pressure from settlement, game control, and poachers sometimes confines huge herds to restricted habitats, which they may destroy. Already three of Tanzania's national parks—Serengeti, Manyara, and Ruaha—have more elephants than is good for them. The elephant problem, where and when and how to manage them, is a great controversy in East Africa, and its solution must affect the balance of animals and man through the continent.

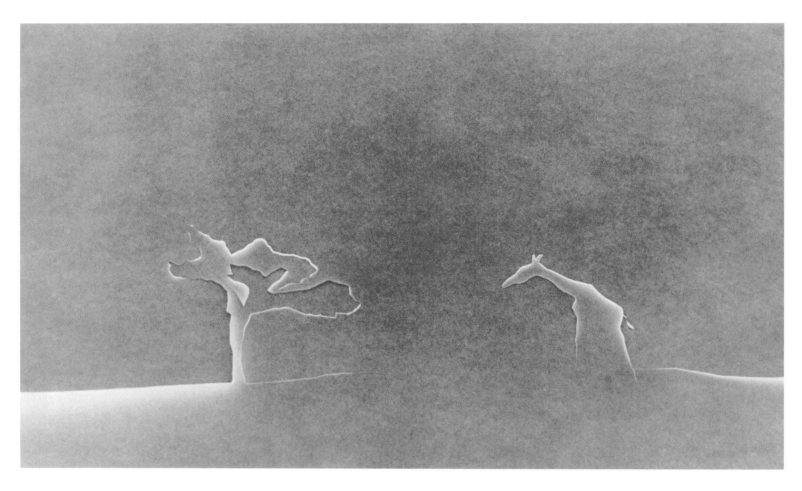

Beneath the plane, the elephant mass moved like gray lava, leaving behind a ruined bog of mud and twisted trees. Anxious to see the great herd from the ground, I picked up George Schaller of the Serengeti Research Institute at Seronera and drove northwest to Banagi, then westward on the Ikoma–Musoma track to the old northwest boundary of the park, where we headed across country. I had taken good bearings from the air, but elephants on the move can go a long way in an hour, and even for a vehicle with four-wheel drive, this rough bush of high grass, potholes, rocks, steep brushy streams, and swampy mud is very different from the hardpan of the plain. The low hot woods lacked rises or landmarks, and for a while it seemed that I had actually misplaced four hundred elephants.

Then six bulls loomed through the trees, lashing the air with their trunks, ears blowing, in stiff-legged swinging stride; they forded a steep gully as the main herd, ahead of them, appeared on a wooded rise. Ranging up and down the gully, we found a place to lurch across, then took off eastward, hoping to find a point downwind of the herd where the elephants would pass. But their pace had slowed as the sun rose; we worked back to them, upwind.

The elephants were destroying a low wood—this is not an exaggeration—with terrible cracking of trees, but after a while they moved out onto open savanna. In a swampy stream they sprayed one another and rolled in the water and coated their hides with mud, filling the air with a thick sloughing sound like the wet-meat sound made by predators on a kill. Even at rest the herd flowed in perpetual motion, the ears like delicate great petals, the ripple of the mud-caked flanks, the coiling trunks—a dream rhythm, a rhythm of wind and trees. "It's a nice life," Schaller said. "Long, and without fear." A young one could be killed by a lion, but only a desperate lion would venture near a herd of elephants, which are among the few creatures that reach old age in the wild.

There has been much testimony to the silent step of elephants, and all of it is true. At one point there came a cracking sound so small that had I not been alert for the stray elephants all around, I might never have seen the mighty bull that bore down on us from behind. A hundred yards away, it came through the scrub and deadwood swift as a cloud shadow. I raised binoculars to watch him turn when he got our scent, but the light wind had shifted and instead the bull was coming fast, looming higher and higher, filling the field of the binoculars, forehead, ears, and back agleam with wet mud dredged up

from the donga. There was no time to reach the car, nothing to do but stand transfixed. A froggish voice said, "What do you think, George?" and got no answer.

Then the bull scented us—the hot wind shifting every moment—and the dark wings flared, filling the sky, and the air was split wide by that ultimate scream that the elephant gives in alarm or agitation, that primordial warped horn note out of oldest Africa. It altered course without missing a stride, not in flight but wary, wide-eared, passing man by. When first aware of us, the bull had been less than one hundred feet away—I walked it off—and he was somewhat nearer where he passed. "He was pretty close," I said finally to Schaller. George cleared his throat. "You don't want them any closer than that," he said. "Not when you're on foot." Schaller, who has no taste for exaggeration, had a very respectful look upon his face.

Stalking the elephants, we were soon a half mile from my Land Rover. What little wind there was continued shifting, and one old cow, getting our scent, flared her ears and lifted her trunk, holding it upraised for a long time like a question mark. There were new calves with the herd, and we went no closer. Then the cow lost the scent, and the sloughing sound resumed, a sound that this same animal has made for 400,000 years. Occasionally there came a brief scream of agitation, or the crack of a killed tree back in the woods, and always a rumbling of elephantine guts, the deepest sound made by any animal on Earth except the whale.

Africa. Noon. The hot still waiting air. A hornbill, gnats, the green hills in the distance, wearing away west toward Lake Victoria. . . .

Dr. Schaller, a lean, intent young man whose work on the mountain gorilla had already made his reputation, was studying the carnivores of the Serengeti. In the winter of 1969, he spent as much time as possible in the field, and often he was kind enough to take me with him. Usually we were under way before the light, when small nocturnal animals were still abroad—the springhares, like enormous gerbils, and the small cats and genets. The eyes of nocturnal animals have a topetum membrane that reflects ambient light, and in the headlights of the Land Rover the eyes of the topi were an eerie silver, and lion eyes were red or green or white, depending on the angle of the light. The night gaze of most animals is red, like a coal-red beacon that we once saw high in the branches of a fever tree over the Seronera River; the single cinder, shooting an impossible distance from one branch to another, was the eye of the lesser gallego, or bushbaby, a primitive small primate that resembles the arboreal creature from which mankind evolved.

In a silver dawn giraffes swayed in the feathery limbs of tall acacias, and a file of warthog trotted away into the early shadows. Where a fork of the river crossed the road, a yellow reedbuck and a waterbuck stood juxtaposed. With its white rump and coarse gray hair, the waterbuck looks like deer, but deer do not occur in Africa south of the Sahara; like the wildebeest, gazelles, and other deerlike ruminants, from the tiny dik-dik to the great cowlike eland, the reedbuck and waterbuck are

antelope, bearing not antlers but hollow horns: the family name, Antilopinae, means "bright-eyed." All antelope share the long ears, large nostrils, and protruding eyes that together with speed help protect them against predators, but nonmigratory species such as topi, waterbuck, and kongoni seem more vigilant than the herd species of the green plain.

At Naabi Hill, the wildebeest were moving east after the rains. In their search for new growth, wildebeest are often seen trooping steadfastly over arid country toward distinct thunderstorms, which bring a flush of green to the parched landscapes. Some 200,000 were in sight at once, with myriad zebra and the small Thomson's gazelle. Eight wild dogs were hunting new gazelle left hidden by their mothers in the tussocks; one snatched a calf out of the grass only yards from the tires of the Land Rover and, with the two nearest dogs, tore it to bits. The death of new calves is quick; they are rended and gone. But one calf older than the rest sprang away before the dog and made a brave run across the plain in stiff-legged long bounces, known as "pronking," in which all four hooves strike the ground at once. Like the electric flickering of the flank stripe, pronking is thought to be a signal of alarm. Though its endurance was astonishing, it lasted so long because most of the dogs were gorged and failed to cooperate. While the lead dog snapped vainly at the flying heels, the rest loitered and gamboled, picking up another calf that one of them ran over in the grass. . . .

The unseasonal rains of late January continued into February, falling in late afternoon and in the night; many days had a high windy sun. In this hard night I walked barefoot on the plain to feel the warm hide of Africa next to my skin, and, aware of my steps, I was also aware of the red oat and red flowers of indigofera, the skulls with the encrusted horns (which are devoured like everything else: a moth lays its eggs upon the horns, and the pupae, encased in crust, feed on the keratin), the lairs of wolf spiders and the white and yellow pierid butterflies like blowing petals, the larks and wheatears (Old English for white-arse), the elliptical hole of the pandanus scorpion and the round hole of the mole cricket, whose mighty song attracts its female from a mile away, the white turd of the bone-eating hyena, and the pyriform egg of a crowned plover in water-colored camouflage that blends the rain and earth and air and grass.

Bright flowers blowing, and small islets of manure; to the manure come shiny scarabs, beloved of Ra (god of the sun, son of the sky). The dung beetles, churning in over the grass, collide with the deposits of a manure, or attach themselves to the slack ungulate stomachs full of half-digested grass that the carnivores have slung aside. They roll neat spheres of ordure larger than themselves and hurry them off over the plain. Here dung beetles fill the role of earthworms: the seasonal droppings of hundreds of thousands of animals, most of which is buried by the beetles, ensures that the soil will be aerated as well as fertilized.

By morning the ground is soaked and tracks muddy. Frogs have sprung from fleeting pools, and the trills of several chorus in the rush to breed. Migrant companies of European storks, nowhere in evidence the day before, come down in slow spirals from towers of the sky to eat frogs, grass mice, and other lowlife that the rains have flooded from the earthen world under the mud-flecked flowers. . . .

At the den one afternoon, four pups were frolicking from dog to dog, and one of the dogs dutifully disgorged some undigested chunks of meat, but the pups were bothered by the bursts of thunder, looking up from their food to whine at the far, silent lightning. A mile to the west, zebra herds moved steadily along the skyline. When the rain thickened, the pups tumbled down into their den, and the adults gathered in a pile of matted black and brindle hair to shed the huge tropical downpour. Toward twilight, the rain eased and the dog pile broke apart, white tail tips twirling; the animals frisked about the soggy plain, greeting anew by inserting the muzzle into the corner of another's mouth, as the pups do when begging food. When they are old enough to follow the hunt, the pups will be given first place at the kill.

Four dogs led by a brindle male moved off a little toward the west; they stood stiff-legged, straining forward, round black ears cocked toward the zebra herds, which, agitated by the storm, were moving along at a steady gallop. Then the four set off at a steady trot, and the others broke their play to watch them go. Three more moved out, though not too swiftly, and George Schaller followed the seven at a little distance. Soon the remaining animals were coming, all but one that remained to guard the den. . . .

The four lead dogs, nearing the herd, broke into an easy run; the zebras spurted. Perhaps the dogs had singled out a victim—an old or diseased horse, a pregnant mare, a foal—for now they were streaming over the wet grass. Rain swept the plain, and gray sheets blurred the swirling stripes, which burst apart to scatter in all directions. The dogs wheeled hard, intent now on a quarry, but lost it as the horses veered, then milled together in a solid phalanx. A stallion charged the dogs, ears back, and they gave way.

The chase of a mile or more had failed; the wild dogs frisked and played. But not all thirteen were intent, and in moments they were loping back past the waiting vehicles, headed west. The leaders were already in their hunting run, bounding along in silence through the growing dark like hounds of hell, and the others, close behind, made sweet puppy-like call

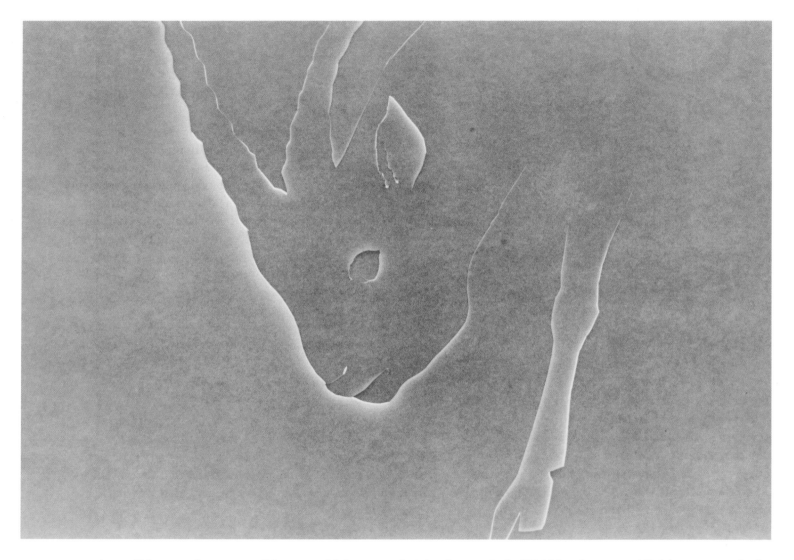

notes, strangely audible over the motor of the car, which swerved violently to miss half-hidden burrows. All thirteen stretched cadaverous shapes in long, easy leaps over the plain. In their run, the dogs were beautiful and swift, and they came up with the herds in not more than a mile. Then the dark shapes were whisking in and out among the zebra, and a foal still brown with a foal's long guard hairs quit the mare's side when the dog bit at it, and was surrounded.

A six-month zebra foal, weighing perhaps three hundred pounds, is too big to be downed easily by thin animals of forty pounds apiece. The dogs chivvied it round and round. One dog had sunk its teeth into the foal's black muzzle, tugging backward to keep the victim's head low—this is a habit of the dogs, to compensate for their light weight—and the dog was swung free of the ground as the foal reared, lost its balance, and went down. Now the mare charged, scattering the pack, and

when the foal jumped up as if unhurt, the two fled for the herd. But the dogs overtook the foal again, snapping at its hams, and, braying softly, it stopped short of its own accord. Again a dog seized its muzzle, legs braced, dragging the head forward, as the rest tore into it from behind and below. At the yanking of its nose, the foal's mouth fell open, and it made a last small sound. Once more the mare rushed at the dogs, and once again, but already she seemed resigned to what was happening, and did not follow up on her own attacks. The foal sank to its knees, neck still stretched by the backing dog, its entrails a dim gleam in the rain. Then the dog at its nose let go and joined the rest, and the foal raised its head, ears high, gazing in silence at the mare, which stood guard over it, motionless. Between her legs, her foal was being eaten alive, and mercifuly, she did nothing. Then the foal sank down, and the dogs surged at the belly, all but one that snapped an eye out as the head flopped on the grass.

Unmarked, the mare turned and walked away. Intent on her foal, the dogs had not once snapped at her. Noticing the car twenty yards off, she gave a snort and a jump sideways, then walked on. Flanks pressed together, ears alert, her band awaited her; nearby, other zebra clans were grazing. Soon the foal's family, carrying the mare with it, moved away, snatching at the grass as they ambled westward.

The foal's spread legs stuck up like sticks from the twisting black and brindle; the dogs drove into the belly, hind legs straining. They snatched a mouthful, gulped it, and tore in again, climbing the carcass, tails erect, as if every lion and hyena on the plain were coming fast to drive them from the kill. All thirteen heads snapped at the meat, so close together that inevitably one yelped, but even when two would worry the same shred, there was never a snarl, only a wet, steady sound of meat-eating. When the first dog moved off, licking its chops, the foal's rib cage was already bare; not ten minutes had passed since it had died. Then the hyenas came. First there were two, rising up out of the raining grass like mud lumps given life. They shambled forward without haste, neither numerous nor hungry enough to drive away the dogs. Then there were five in a semicircle, feinting a little. A dog ran out to chase away the boldest, and then two of the five, with the strange speed that makes them deadly hunters on their own, chased off a sixth hyena—not a clan member, apparently—that had come in from the north through the twilight rain.

Six of the dogs, their feeding finished, wagged long tails as they romped and greeted; there was just enough light to illuminate the red on their white patches. The rest fed steadily, eyes turned to the hyenas as they swallowed, and as each dog got its fill and forsook the carcass, the half circle of hyenas tightened. The last dog gave way to them without a snarl. The forequarters were left, and the head and neck, and all the bones. For the powerful jaws of the hyena, the bones of the plains animals present no problem. Hoofs, bones, and skin of what had been, ten minutes before, the fat hindquarters of a swift, skittish young horse lay twisted up in a torn muddy bag; the teeth of its skull and the white eye sockets were luminous. At dark, as the tail tips of the dogs danced away eastward, the hyena shapes drew together at the remains, one great night beast sinking slowly into the mud. . . .

I once watched a hyena gaining on a wildebeest that only saved itself by plunging into the heart of a panicked herd; the hyena lost track of its quarry when the herd stampeded. The cringing bearlike lope of these strange cat relatives is deceptive: a hyena can run forty miles an hour, which is considered the top speed of the swift wild dog. Cheetah are said to attain sixty but have small endurance; I have seen one spring at a Thompson's gazelle, its usual prey, and quit within the first one hundred yards. Hyenas, on the other hand, will run their prey into the ground; there is no escape. And in darkness they are bold—a man alone on the night roads of Africa has less to fear from lions than hyena. In the Ngorongoro Crater the roles ordinarily assigned to lions and hyenas are reversed. The hyenas, hunting at night, make most of the kills, and the lions seen on the carcasses in daytime are the scavengers. The crater's hyenas are divided into great clans, and sometimes these hyena armies war at night, filling the crater with the din of the inferno.

The natural history of even the best-known African mammals is incomplete, and such hole dwellers as the ant-bear, the aard-wolf, and the pangolin have avoided the scrutiny of man almost entirely. It is not even known which species excavate the holes, which may also be occupied by hyenas, jackals, mongooses, bat-eared foxes and wolves, porcupines, ratels or honey badgers, and, in whelping season, wild hunting dogs. Often the burrows are dug in the bases of old termite hills, which stand on the plain like strange red statues of a vanished civilization, worn to anonymity by time. The termites are ancient relatives of the cockroach, and in the wake of rains they leave the termitaria in nuptial flight; soon their wings break off, and new colonies are founded where they come to earth. Were man to destroy the many creatures that prey on them, the termite mounds would cover entire landscapes. The African past lies in the belly of the termite, which has eaten all trace of past tropical civilizations and will do as much for the greater part of what now stands. . . .

Hippos can weigh twice as much as buffalo, or a ton and a half each. Like whales, they are born in the water, but ordinarily they feed on land, and their copious manure, supporting rich growth of blue-green algae, is a great boon to fish. At the Mara River they had piled up in the rapids, where the lateritic silt had turned them the same red as the broad-backed boulders. On land, the hippopotamus exudes a red secretion, perhaps to protect its skin against the sun, and Africans say that it is sweating its own blood. With their flayed skins, cavernous raw mouths, and bulging eyes, their tuba voices spitting the wash of the Mara on its banks seemed like the uproar of the damned, as if, in the cold rain and purgatorial din, just at this moment, the great water pigs had been cast into perdition, their downfall heralded by the scream of the fish eagle that circled overhead. . . .

In March, renewal of the southeast monsoon brings the long rains. A somber light refracted from the water gleamed in the depressions, and the treeless distances with their animal silhouettes, the glow of bright flowers underfoot, recalled the tundras of the north to which the migrant plovers on the plain would soon return.

The animals had slowed, and some stood still. In this light those without movement looked enormous, the archetypal animal cast in stone. The ostrich is huge on the horizon, and the kori bustard is the heaviest of all flying birds on Earth. Everywhere the clouds were crossed by giant birds in their slow circles like winged reptiles on an antediluvian sky.

Mid-March, when the long rains were due, was a time of wind and dry days in the Serengeti, with black trees in iron silhouette on the hard sunsets and great birds turning forever on a silver sky. A full moon rose in a night rainbow, but the next day the sun was clear again, flat as a disk in the pale universe.

Two rhinos and a herd of buffalo had brought up the rear of the eastward migration. Unlike the antelope, which blow with the wind and grasses, the dark animals stood earthbound on the plain. The antelope, all but a few, had drifted east under the Crater Highlands, whereas the zebra, in expectation of the rains, were turning west again toward the woods. Great herds had gathered at the Seronera River, where the local prides of lions were well fed. Twenty lions together, dozing in the golden grass, could sometimes be located by the wave of a black tail tuft or the black ear tips of a lifted head that gazed through the sun shimmer of the seed heads. Others gorged in uproar near the river crossing, tearing the fat striped flanks on fresh green beds—now daytime kills were common. Yet for all their prosperity, there was an air of doom about the lions. The males, especially, seemed too big, and they walked too slowly between feast and famine, as if in some dim intuition that the time of the great predators was running out.

Pairs of male lions, unattached to any pride, may hunt and live together in great harmony, with something like demonstrative affection. But when two strangers meet, there seems to be a waiting period while fear settles. One sinks into the grass at a little distance, and for a long time they watch each other, and their sad eyes, unblinking, never move. The gaze is the warning, and it is the same gaze, wary but unwavering, with which lions confront man. The gold cat eyes shimmer with hidden lights, eyes that see everything and betray nothing. When the lion is satisfied that the threat is past, the head is turned, as if ignoring it might speed the departure of an unwelcome and evil-smelling presence. In its torpor and detachment, the lion sometimes seems the dullest beast in Africa, but one has only to watch a file of lions setting off on the evening hunt to be awed anew by the power of this animal. . . .

One late afternoon of March, beyond Maasai Kopjes, eleven lionesses lay on a kill, and the upraised heads, in a setting sun, were red. With their grim visages and flat gazed eyes, these twilight beasts were ominous. Then the gory heads all turned as one, ear tips alert. No animal was in sight, and their bellies were full, yet they glared steadfastly away into the emptiness of plain, as if something that no man could sense was imminent.

Not far off there was a leopard; possibly they scented it. The leopard lay on an open rise, in the shadow of wind-worn bush, and unlike the lions, it lay gracefully. Even stretched on a tree limb, all four feet hanging, as it is seen sometimes in the fever trees, the leopard has the grace of complete awareness, with all its tensions in its pointed eyes. The lion's gaze is merely baleful; that of the leopard is malevolent, a distillation of the trapped fear that is true savagery.

Under a whistling thorn the leopard lay, gold coat on fire in the sinking sun, as if imagining that so long as it lay still it was unseen. Behind it was a solitary thorn tree, black and bony in the sunset, and from a crotch in a high branch, turning gently, torn hide matted with caked blood, the hollow form of a gazelle hung by the neck. At the insistence of the wind, the delicate black shells of the turning hoofs, on tiptoe, made a dry clicking in the silence of the plain.

Gol Mountains

The lonely Gols are kopjes, or granite outcrops, of the short-grass prairie, lying off the north of the Olduval-Seronera road. The shy cheetah is a creature of the Gols, its gaunt gait and sere pale coat well suited to these wind-withered stones, and one day a glowing leopard stretched full length on a rim of rock, in the flickering sun shade of a fig; seeing man, it gathered itself without stirring, and flattened into the stone as it slid from view. It is said that a leopard will lie silent even when struck by stones hurled at its hiding place—an act that would bring a charge from any lion—but should its burning gaze be met, and it realizes it has been seen, it will charge at once. A big leopard is small by comparison to a lion, or to most men for that matter, but its hind claws, raking downward, can gut its prey even as the jaws lock on the throat, and it strikes fast. It is one of the rare creatures besides man known to kill for

the sake of killing and, cornered, can be as dangerous as any animal in Africa. The strength of leopards is intense: I have seen one descend the trunk of a tall tree headfirst, a full-grown gazelle between its jaws. . . .

There are few trees in the Gols, which are low and barren yet in their way as stirring as the Morus, which rise like monuments in a parkland of twenty-five square miles and have a heavy vegetation. Impala, buffalo, and elephant are attracted to the Morus from the western woods, and the elephants, which are celebrated climbers, attain the crests of the steep kopjes, to judge from the evidence heaped on the rock. One day at noon, from this elephant crest, a leopard could be seen on the stone face of the kopje to the south, crossing the skeletal shadows of a huge candelabra euphorbia. In the stillness that attended the cat's passage, the only sound was a rattling of termites in the leaf litter beneath my feet.

Here near the woods the long grass is avoided by the herds. Lone topi and kongoni wander among these towers, and an acacia is ringed by a bright circlet of zebras, tails swishing, heavy heads alert. The wild horses are not alarmed by man, not yet; all face in another direction. Somewhere upwind, in the tawny grass, there is a lion. . . .

A big weather wind from the southeast came up in the late morning, and by noon it was shifting to the east, turning the dark clouds on the Crater Highlands. Once again, the herds were moving. To see the animals in storm, I went over to the great rock called Soit Naado Murt, and climbed the broad open boulders on the south side, away from the road.

Soaring thunderheads, unholy light: at the summit of the rock the wind flung black leaves of twining fig trees flat against the sky, and black ravens blew among them. I straightened, taking a deep breath. From its aerie, a dog baboon reviled me with fear and fury. Puffs of cold air and a high far silent lightning; thunder rolled up and down the sky. Everywhere westward, the zebra legions fled across the plain. But dark was coming, and soon I hurried down off the high places. At the base of the rock, the suspense, the malevolence in the heavy air was shattered by a crash in the brush behind, and I whirled toward the two tawny forms that hurtled outward in a bad light, sure that this split second was my last. But the lion-size and lion-colored animals were a pair of reedbuck, as frightened as myself, that veered away toward the dense cover of the korongo. I stood still for a long time, staring after them as darkness fell, aware of a strange screaming in my ears. Then I came to, and moved away from the shadows of the rock. A pale band in the west, under mountains of black rain, was the last light, and against this light, on the rock pinnacles, rose the hostile cliff tribe of baboons. In silhouette they looked like early hominids, hurling wild manic howling at my head. . . .

A day of low still clouds, waxy gray with the weight of rain. . . . At Naabi Hill, the eastern portal of the Serengeti Park, three lionesses lay torpid on a zebra. Vultures nodded in the low acacias, and the hyenas, wet hair matted with filth on their sagging bellies, dragged themselves, tails tight between their legs, from the rain wallows in the road. We turned off north toward Loliondo, then east again under Lemuta Hill. Between Lemuta and the Gols is a dry valley, in the rain shadow of the Crater Highlands; here the hollow calls of sand grouse resound in the still air, and an echo of wind from the stoop of a bateleur eagle. Where we had come from and where we were going, a pale green softened the short grass, but in the shadow of the rain, despite massed clouds in all the distances, the flicked hoof of a gazelle raised the soil in a spiral of thin dust.

Lightning came, and a drum of rain on the hard ground across the valley. On the taut skin of Africa rain can be heard two miles away. Birdcalls rang against Naisera's walls, which on the north are painted white by high nests of hawk and raven. At the summit, in the changing light, the swifts and kestrels swooped and curved, and an Egyptian vulture gathered light in its white wings. Then the sky rumbled and the white bird sailed on its shaft of sun into the thunder. . . .

Like many white men that one comes across in Africa, Myles Turner is a solitary person whose job as park warden of the Serengeti keeps him in touch with mankind more than he would like, but one day he got away on a short safari and was kind enough to take me with him. We would go to the Gol Mountains, in Maasai Land, and from there attempt to reach by Land Rover that part of the Rift Escarpment that stands opposite a remote volcano known to the Maasai as *ol doinyo le eng ai*, the Mountain of the God, commonly called Lengai.

Ol Doinyo Lengai, though shrouded, was a heavy presence in the sky. Lost bands of kongoni and gazelle wandered down out of the highland clouds and waited for everything and nothing on the edges of the cinder plain. On a rise stood a

stone oryx with one horn. The horn was long and straight and whorled; here was the unicorn. The beisa oryx is a strong gray antelope, wary and quick and spirited; oryx have been known to kill attacking lions. This one, given sudden life, went off at a fast trot. Far down the slope its herds were already moving at our approach. Myles said that in this region, where the animal is rare and wild, the hunter who killed an oryx usually earned it.

On the level ground, the game trails radiated out in cracks from the dry waterholes. Near one hole a dead zebra, still intact, had drawn a horde of griffons from the sky. The zebra did not look diseased, and we searched it for signs of Sonjo arrows, but there was nothing; it had merely died. . . . At midday, Lengai loomed through the clouds and again vanished. Heat and silence became one. Adding their silence to the silence, the griffons waited.

Round-haunched zebra stood, tails blowing, on a round curve of a hill, each horse in silhouette against the sky. A cheetah appeared, and then two more, moving westward up the valley; the animal survives in such dry country by lapping blood from the body cavity of its kill. The walk of lions is low-slung and easy, and leopards move like snakes, striking and coiling; the cheetah's walk looks stiff and deadly as if it were bent upon revenge. The three cats were traveling, not hunting, and did not look back.

On the plain lay a tawny pipit, dead, raked by a hawk. Somewhere jackals were keening, and a restless lion roared from across the valley. Then owls emerged, and in the growing dark, the white bellies of gazelles flashed back and forth like flanges in a ghostly dance. From the mouth of a burrow peered four faces of bat-eared foxes, and from Naisera came a troop of mongoose, looping out in single file over the plain; it was the time of the night hunters.

Barafu Kopjes

At dawn we left the Gols behind, turning north toward Loliondo. A wind from the northeast was high and cold, a wind of hawks and eagles, and beyond Lemuta the delicate pearl-and-chestnut kestrels dipped and rose, snatching dung beetles from the hard-caked ground. Farther on, four steppe eagles in a half circle at the mouth of a hole fed with ravenous dexterity on hatching termites. Africa is a place of incongruities, as if its species were still evolving—kingfishers that live in the dry woods, owls that seize fish, eagles that eat insects. And doubtless the great variety of raptors here is accounted for by their versatility of habit: nothing is overlooked, and nothing is wasted.

Across the plain came a strange hyena that behaved like no hyena we had ever seen. Though unpursued, and pursuing no other creature, it ran hard, and though its head was half-averted in the manner of hyenas, its tail was raised, not tucked in the usual way between the legs, and it came straight for the car instead of fleeing it, only turning off in final yards, still unafraid, still searching.

The vehicle traversed the lonely rises, rolling a thin dust cloud toward the west. Myles wished to show me an enormous fig that stands by itself far off beyond Barafu Kopjes. These hard plains are bare and bony, with only a whisper of grass, yet the animals keep to the ridges, where the grass is shortest. In a tilted world, the wildebeest went streaming down the sky, black tail tassels hung on the wind behind, all but a solitary bull, thin-ribbed and rag-tailed, old beard blowing. Perhaps he felt his death upon him, for he paid no attention to our intrusion. Soon he had the whole sky to himself.

The giant fig, which looks like a small grove in the distance, is at least as old as man's recorded history on this plain. Its spread is not less than 150 feet, the size of six ordinary figs, and it is a tree of life. Cape rooks, kestrels, owls, and the shy brown-chested cuckoo were in residence, and none would willingly leave the tree because there are no other trees for miles around. One owl that moved onto a nearby rock was punished by the kestrels; at each blow from above, it shifted its feet and shuffled its loose feathers.

The tree has a Maasai hearth built into its thick base, and a flat stone near at hand for sharpening spear blades. One day I would like to sit under this tree that has drawn so much fat wood and fleshy leaves out of near-desert, and stare for a week or more into the emptiness. One understands why these monumental figs take on a religious aura for the Africans; they are thought to symbolize the sacred mountains, and the old ways of close kinship with the earth and rain, Nature and God.

The great fig west of the Gol Mountains overlooks a dry korongo, and nearby there is a Maasai cattle well. In the well lay a drowned hyena so blue and bloated that the rotting skin shone through the wide-stretched hairs. Though it had been there many days, no scavenger had touched it. Even its eye was still in place, fixed malevolently upon the heavens.

We headed south. Miles from where it had first appeared, the lone hyena rose out of the land, and this time it came even closer, loping along beside the car, tail high and bald eye searching. We wondered then if this haunted beast was hunting for its mate, and if the mate might be the hyena in the well. But we did not know, and never would, and the mystery pleased us.

Lake Manyara

Lake Manyara is a soda lake, or *magadi*, that lies along the base of the Rift Escarpment. The east side of the lake lies in arid plain, but the west shore, where streams emerge from the porous volcanic rock of the Crater Highlands, supports high, dark groundwater forest. The thick trees have the atmosphere of jungle, but there are no epiphytes or mosses, for the air is dry. On the road south into Lake Manyara Park, this forest gives way to open wood of that airiest of all acacias, the umbrella thorn, and beyond the Ndala River is a region of dense thicket and wet savanna. The strip of trees between lake and escarpment is so narrow, and the pressure on elephants in the surrounding country so great, that Manyara can claim the greatest elephant concentration in East Africa, an estimated twelve to the square mile.

In acacia wood that descends to the lakeshore, elephants were everywhere in groves and thickets. Elephants travel in matriarchal groups led by a succession of mothers and daughters—female elephants stay with their mothers all their lives— and this group may include young males that have not yet been driven off. (Elephants not fully grown are difficult to sex—their genitals are well camouflaged in the cascade of slack and wrinkles—and unless their behavior has been studied for some time, the exact composition of a cow-calf group is very difficult to determine.) Ordinarily the leader is the oldest cow, who is related to every other animal; she may be fifty years old and past the breeding age, but her great memory and experience are the herd's defense against drought and flood and man. She knows not only where good browse may be found in different seasons, but also when to charge and when to flee, and it is to her that the herd turns in time of stress. When a cow is in season, bulls may join the cow-calf group; at other times, they live alone or in herds of bachelors. When I drove near, the bulls moved off after a perfunctory threat display—flared ears, brandished tusks, a swaying forefoot like a pendulum, the dismantling of the nearest tree, and perhaps a diffident scream; sometimes they ease their nervous strain by chasing a jackal or a bird. With cows, as well, aggressive behavior is usually mere threat display, though it is wiser not to count on it.

When I first came to Lake Manyara, in the winter drought of 1961, it was a pale dead place of scum froth and cracked soda; in 1969 and 1970 the water level was so high that many of the tracks behind the shore were underwater, and

much of the umbrella thorn was killed. At twilight one late afternoon near the drowned forest, a herd of elephants fed on mats of dead typha sedge blown over from the far side of the lake. The animals waded to their chests in the greasy waves, trunks coiling in and out, ears blowing. Night was falling in the shadows of the Rift, which rose in a black wall behind the elephants, and from dusky woods came a solitary fluting. As the sun sank to the escarpment, the western sky took on a greenish cast, and the last light of storm caught the whiskers on the pointed lips, the torn flutings of the ears, the ragged switch of a wet tail on ancient hide. The dead forest, the doomed giants, the wild light were of another age, and made me restless, as if awakening ancestral memories of the Deluge.

Lengai

Broad-backed, motionless on the wind, an eagle descended the black river that isolates Ol Doinyo Lengai from the Crater Highlands. Seen from above, a bird of prey, intent on all beneath, is the very messenger of silence.

A series of small mounds, like stepping-stones, emerged from the smooth surface of the ash; half-blind with effort on the climb, I had scarcely noticed them. The mounds formed a distinct line down the crest of the ridge, like rhinoceros prints elevated above the surface, and as it happens this was what they were. Apparently a rhino high up on the mountain had tried to flee the last eruptions—perhaps in vain, since its tracks vanish near the edge of the ravine. There was no sign of a trail leading upward, only down. Its tread had compacted the hot ash, and afterward the mountain winds had worn away the uncompacted ash all around until the prints had risen above the surface.

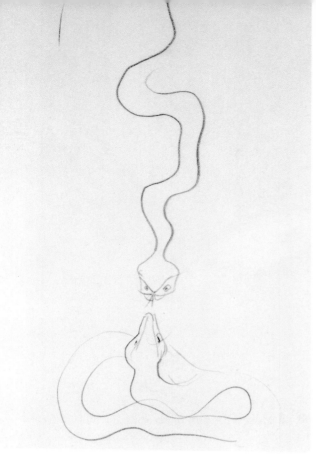

Holding a hoofprint in my hands, I raised my eyes to where that horned lump, as if spat up by the volcano, had taken form in the poisonous clouds and rushed down the fiery ridge. What had drawn it up into the mists? Had it been blind, like the buffalo found in the snow high on Mount Kenya? Imagine the sight of that dark thing in the smoke of the volcano; had an African seen it, the rhino might have become a beast of legend, like the hyena, for it is in such dreamlike events that myths are born.

Anxious to transfix so great a mystery, I chipped two prints clean of packed ash and wedged them into my pack. We descended the volcano, crossed the ash plains, circled dust storms. For four hours on sore feet, I carried the stone prints, but they belonged to the dead mountain, for in the journey they returned to dust.

Mount Meru

The buffalo is said to be the most aggressive animal in Africa, much more dangerous than the rhino, since that beast will often thunder past its target and keep right on going until, at some point in its course, having met with no obstacle and having forgotten what excited it in the first place, it comes to a ponderous halt. The buffalo, on the other hand, turns quickly and is diligent in its pursuit. It is keen of nose and eye and ear, and like the lion, is very difficult to stop once it attacks, often persisting in the work of destruction for some time after the object of its rage or fear is dead. It will even stalk a man, especially when wounded, coming around on its pursuer from behind.

A buffalo skull, as if in sign, lay in the grass, surrounded by fresh spoor. We [Peter Matthiessen, botanist Desmond Vesey-Fitzgerald, orinthologist John Beasley, and Meru rangers] stood and listened. Before making their move, buffalo may lie in wait until whatever approaches them has gone past. (This is customarily ascribed to malevolence or low cunning, but dull wits and slow reaction time may be an alternative explanation.) Then the suspense got to the buffalo, and the hidden herd rushed away down the mountainside with a heavy cracking and a long rumbling like a mountain torrent past big boulders. Immediately, a voice called out, *"Kifaru!"* and Vesey mopped his brow. The Meru were pointing at a rhino print as fresh as a black petal, and within seconds a rhino crashed into the brush off the east. The crash started up a buffalo lying low in the wild nightshade to our left. This lone animal was the one we were afraid of, and as it was much closer than the others, its explosion through the branches caused the Wazungu [the whites] to rush in all directions. Beasley sprinted past, bound for the same tree as myself, as the askari Serekieli, standing fast, fired his gun to turn the charge. The black blur whirled away, and the echo died.

The great trees, fallen, have opened glades in a wild parkland, and silver deadwood is entwined by a climbing acantha with blossoms of light lavender. In stillness, in wind-shifted sun and shadow, a papilio butterfly, deep blue, is dancing with a Meru swallowtail, black and white, which ascends from the black and white remains of a colobus monkey, knocked from its tree by a leopard or an eagle. . . .

"Kifaru!"

At six thousand feet, in a mahogany grove, a rhino digging is so fresh that it seems to breathe; we hurry past. "I must say," Vesey huffs, "on leave in England, it's nice to walk about a bit and not have some bloody ungrateful beast rush you." Once Vesey was chased by an irate hippopotamus that took a colossal bite out of his Land Rover.

A shy lemon dove in a pepper tree . . . more spoor . . .

"Kifaru!"

At the crash, we scatter. Horn high, tail high, a rhino lumbers forth out of the undergrowth thirty yards away. The rhino is said to hoist its tail when wishing to depart, but no one appears confident that this is so. From behind my tree, too big to climb, I see Beasley on the limb of a wild coffee, with Vesey crouched below, as the rhino, trotting heavily across the glade, emits three horrible coughing snorts. The askari Frank is somewhere out of sight, but Serekieli stands bravely, legs apart, ready to fire. There is no need—the rhino goes, and keeps on going.

The Africans permit themselves a wild sweet laughter of relief, watching the whites come down out of the trees. Vesey, treed twice in half an hour, is not amused.

"Get on! Get on!" he says, anxious to be off this bloody mountain.

Ngurdoto Crater

I hoped to see the white-maned Kilimanjaro bushpig, and one afternoon I went down into Ngurdoto Crater with Serekieli. . . . Unlike Ngorongoro, there are no tracks or paths inside the crater, and as one peers down from the rim at remote creatures grazing in peace, oblivious of man, there rises from this hidden world that stillness of the early morning before man was born.

An elephant path of pressed humus and round leather-polished stones wound down among the boles of the gallery forest. The sun was high and the birds still, the forest dark and cool. Under the steep rim, out of the reach of axes, rose the great African mahogany and the elegant tropical olive, *loliondo*. The steep path leveled out into grassy glades, which, being ponds in time of rain, are mostly round. We followed them eastward, under the crater rim, working our way out of the trees. Baboon and warthog stared, then ran, the hominoids barking and shrieking as they scampered, the wild pigs departing the field in a stiff trot. Moments later we stood exposed in a bowl of sunlight.

Buffalo and a solitary rhino took mute note of us; the world stood still. Flat wet dung raised its reassuring smell in halos of loud flies. We turned west across wild pasture—cropped turf, cabbage butterflies, and cloven prints filled with clear rain—that rings the sedge swamp in the pit of the caldera. A hawk rose on thermals from the crater floor, and white egrets crossed the dark walls; in the marsh, a golden sedge was seeding in the swelling light of afternoon. More buffalo lay along the wood edge at the western wall, and with them rhinos and an elephant. The rhinos lay still, but the elephant, a mile away, blared in alarm, and others answered from the galleries of trees, the screams echoing around the crater; the elephant's ears flared wide and closed as it passed with saintly tread into the forest. Bushbuck and waterbuck lifted carved heads to watch man's coming; their tails switched and their hind legs stamped but they did not run. Perhaps the white-maned bushpig saw us, too, raising red eyes from the snuffled dirt and scratching its raspy hide with a sharp hoof, but today it remained hidden.

The buffalo rose and split into two companies, and twelve hundred hoofs thundered at once under the walls. The thunder set off an insane screeching of baboons that spread the length and breadth of Ngurdoto, and a blue monkey dropped from a lone tree in the savanna and scampered to the forest. Some of the milling buffalo plunged off into the wood, but others turned and came straight at us, the sunlight spinning on their horns. Buffalo have good eyesight, and we expected these to veer, but a hundred yards away, they were still coming, rocking heavily across the meadow. We turned and ran. Confused by our flight, they wheeled about and fled after the rest into the thickets. There came a terrific crack and crashing, as if their companions had turned back and the two groups had collided. In the stunned silence, we headed once more for the western wall but were scarcely in the clear when the rumbling increased again, and the wood edge quaked, swayed, and split wide as the tide of buffalo broke free onto the plain and scattered in all directions.

The hawk, clearing the crater rim, was burned black by the western sun. From the forest, the hollow laugh of the blue monkey was answered by the froggish racketing of a turaco. Parting leaves with long shy fingers, Serekieli probed for sign of an animal trail that might climb to the western rim. We pushed through heavy growth of sage and psidia, stopping each moment to listen hard, then clap our hands. More than an hour was required to climb out of the heat and thicket to the gallery forest under the crater rim, and all the while the elephants were near, in enormous silence.

The leaves hung still. Bright on the dark humus lay a fiery fruit, white bird droppings, the blood-red feather of a turaco. When, near at hand, an elephant blared, the threat ricocheted around the walls, counterpointed by weird echoes of baboons. Serekieli offered an innocent smile and moved quietly ahead. The wood smell was infused with scents of the wild orange and wild pepper trees, and of *Tabernae montana,* a white-flowered relative of frangipani. Where the western sun illumined the high leaves, a company of colobus and blue monkeys, silhouetted, leapt into the sky, careening down onto the canopy of the crater's outer wall. Somewhere elephants were moving. It was near evening, and in every part the forest creaked with life.

Summer 1970

Yaida Valley

One winter day, returning to Seronera from Arusha, Myles Turner flew around the south side of the Crater Highlands, which lay hidden in its black tumulus of clouds. The light plane skirted Lake Manyara and the dusty flats of the witch-ridden Mbugwe, then crossed Mbulu Land, on the Kainam Plateau. Soon it passed over a great silent valley. "That's the Yaida," Turner told me. "That's where those Bushman people are, the Wa-Tindiga." Down there in that arid and inhospitable stillness, cut off from a changing Africa by the ramparts of the Rift, last bands of the Old People turned their heads toward the hard silver bird that crossed their sky. There was no smoke, no village to be seen nor any sign of man.

Later that winter, at Ndala, the elephant biologist Iain Douglas-Hamilton had suggested a safari to Tindiga Land, where his friend Peter Enderlein had lived alone for several years, and was in touch with wild Tindiga still living in the bush.

But Iain was never able to get away, and a year had passed before I crossed paths with Enderlein in Arusha, and arranged to visit his Yaida Chini game post in the summer. . . .

Off among black wrinkled trunks and silvered thorns Giga has glimpsed a shift of shadows. It is a hunting party, crouching low in the golden grass to peer under the limbs; the black of their skin is the old black of acacia bark in shadow. Giga is smiling at them, and they do not run; they have seen Giga, and they have a fresh-killed zebra. Enderlein is grinning freely for the first time since I have known him. "Oh, we are *lucky!*" he says twice; he had not thought we would find the hunters this first day.

A striped hock shines in the fork of a tree; the rest rides on the hunters' shoulders. There are ten Hadza, seven with bows and three young boys, and all are smiling. Each boy has glistening raw meat slung over his shoulders and wrapped around him, and one wears the striped hide outward, in a vest. Except for beads at neck and waist, the boys are naked. The men wear loincloths faded to an olive-earth color that blends with the tawny grass; the rags are bound at the waist by a hide thong, and some have simple necklaces of red-and-yellow berry-colored beads. All wear crude sheath knives in the center of the back, and one has a guinea-fowl feather in his hair.

Shy, they await in a half circle, much less tall than their bows. *"Tsifiaqua mtana,"* and then the hunters say, *"M-taaa-na!"* for warm emphasis, smiling wholeheartedly. (*Tsifiaqua* is "afternoon," as in "Good afternoon," and *mtana* is "nice," as in "Nice day," and *Tsifiaqua m-taa-na*, as the hunters say it, may mean "Oh beautiful day!") I am smiling wholeheartedly too, and so is Enderlein; my smile seems to travel right around my head. The encounter in the sunny wood is much too simple, too beautiful to be real, yet it is more real than anything I have known in a long time. I feel a warm flood of relief, as if I had been away all my life and had come home again—I want to embrace them all. And so both groups stand face to face, admiring each other in the sunlight, and then hands are taken all around, each man being greeted separately by all the rest. They are happy we are to visit them and delighted to pile the zebra meat into the Land Rover, for the day is hot and dry and from here to where these Hadza live, behind the Sipunga Hills, is perhaps six miles of stony walking.

At Gidabembe Hill, among the monoliths, baboons are raving, and there comes a sudden brief strange sound that brings Giga from his cave. *"Chui,"* he whispers. Leopard. But the others shrug—how can one know? The Hadza never like to give opinions. A few days later, in this place, we find the vulture-gutted body of a young leopard on an open slope where no sick leopard would ever lie, and the grass all about has been bent and stamped by a convocation of baboons, as if the creature had been caught in the open by the huge baboon troop, which had killed it. Yet there was no baboon fur in its mouth, nor any blood or sign of struggle in the grass.

The dark falls quiet once again. From Sipunga comes the night song of the unknown birds, and the shrill ringing yip of distant jackals, and inevitably the ululation of hyenas. The Hadza are comparatively unsuperstitious, and unfrightened of the dark: "We are ready for him," they say of Fisi [the hyena], reaching out to touch their bows.

In the shadows ten yards to the left, the cave of leaves is filled with a massive shape, as still as stone. A little way back there was fresh rhino track, and Peter thinks this is the rhino. He circles out a little ways, just to make sure. A slight movement may bring on a rhino charge—its poor vision cannot make out what's moving, and its poor nerves cannot tolerate suspense— whereas a sudden movement may put it to flight. I am considering a sudden movement, such as flight of my own, when I see a tail in a thin shaft of light, and the tail tuft's fleeting silhouette, and grunt at Peter, "Buffalo."

A sun glint on the moisture at the nostril; the animal is facing us. The tail does not move again. We stand there for long seconds, at a loss. Enderlein cannot get a fair shot in the poor light, and at such close quarters, he does not want a wounded

buffalo. He starts a wide circling stalk of the entire copse, signaling his game scouts to follow. But it is the boy Mugunga who jumps forward, and the game scouts shrug, content to let him go. We follow carefully, but soon the hunters vanish in the bushes. Heat and silence. Soon the silence is intensified by a shy birdsong, incomplete, like a child's question gone unanswered.

The bird sings again, waits, sings again. Bees come and go. Soon Mugunga reappears. The beast will not be chivvied out of hiding, and there is no hope of a clear shot with the rifle. But poisoned arrow need not be precise. The hunter had only to wait a few hours before tracking, so as not to drive the dying animal too far away, and in this time he would return to camp to find help in cutting up the meat, or if the animal was big, to move the whole camp to the carcass.

Mugunga draws on Yaida's bow, then picks the stronger bow of Salibogo. The Hadza faces fill with joy; they respect the rifle but they trust the bow. Then Mugunga vanishes once more, and the silence deepens. Leaves stir and are still.

The birdsong ceases as the buffalo crashes free, but there is no shout, no rifle shot, only more silence. Even so we will track this buffalo; Peter keeps the gun. The Hadza move on, bush by bush, glade by glade, checking grass, earth, and twigs, darting through copses where one would have thought so large an animal could not have gone. To watch such tracking is a pleasure, but this is taut work, for the buffalo is listening, it has not taken flight. Somewhere in the silent trees, the dark animal is standing still, or circling to come up behind. Wherever it is, it is too close.

In the growing heat, our nerves go dead, and we are pushing stupidly ahead, inattentive, not alert. When the spoor dies, we cut away from the river in search of another animal. But the sun is climbing, and the big animals will have taken to the shade. The chance of catching one still grazing in the open is now small.

In a swampy place the Hadza fall on a tomato bush. The small fruits are warm red, intensely flavored, and we eat what we can and tie the rest into a rag to bring back to Gidabembe. Not that the hunters feel obliged to do this: men and women seek and eat food separately and quickly, to avoid the bad manners of refusing it to others, and occasional sharing between the sexes is a matter of whim. Farther on, Yaida and Salibogo locate honey in a tree, and again the hunt for buffalo is abandoned. Usually a grass torch is stuck into the hole to smoke out the bees, but the Hadza are more casual than most Africans about bee stings, and Yaida is wringing one stung hand while feeding himself with the other. The honeycomb is eaten quickly, wax, larvae, and all. . . .

We climb steadily through the early morning, across dry open hillsides without flowers. In a broad pile of dik-dik droppings on the trail is a small hole six inches deep and six across. Though it moves in daylight with the shadows of rock and bush, the tiny antelope returns at night to these rabbit heaps out in the open; here it feels safe from stealing enemies, and waits out the long African dusk. Dik-dik (so the Dorobo say) once tripped over the mighty pile of an elephant, and has tried ever since to reply in kind by collecting its tiny droppings in one place. Man takes advantage of the habit by concealing in a hole a ring of thorns with the points facing inward and down. The dik-dik—meaning "quick-quick" in Swahili—cannot extract its delicate leg, and is killed by the first predator to come along. Whoever is hunting here is not a Hadza, for the Hadza know nothing of traps or snares of any kind.

Black rhinos, also sedentary in their habits, follow the same trails to water, dust wallow, and browse, and on a grand scale share this custom of adding to old piles of their own droppings, which are then booted all about, perhaps as a means of marking territory but more likely as an aid to orientation in a beast whose prodigious sniff must compensate for its poor eyesight. Rhino piles are common on this path, together with wallows and the primitive three-toed print. Not far away, one or more of these beasts is listening, flicking its ears separately in the adaptation that accounts in part for its uncanny hearing, and making up its rudimentary mind whether or not to clear the air with a healthy charge.

Southeastward, under the soaring rock, we follow in the noble paths of elephant. Maduru points at an overhanging wall, like a wave of granite of the yellow sky: Darashagan. A hot climb brings us out at last onto a ledge under the overhang, well-hidden by the tops of trees that rise from the slopes below; the ledge looks south down the whole length of the Yaida Valley. There is a hearth here, still in use, and on the wall behind the hearth, sheltered by the overhang, are strong paintings in a faded red of a buffalo and a giraffe. We stand before them in a line, in respectful silence. One day another man, all nerves and blood and hope just like ourselves, drew these emblems of existence with a sharpened bird-bone spatula, a twist of fur, a feather, and others squatted here to watch, much as the Hadza are squatting now. The Mbulu and Barabaig have no tradition of rock painting, whereas the Bushmen, before they became fugitives, made paintings very similar to these.

Andaranda makes a fire and broils hyrax and guinea fowl. When we have eaten, he picks grewia leaves, and the Hadza trim the leaves and roll tobacco from their pouches. I try Nangal's uncultivated weed, and the Hadza giggle at my coughs. Of the drawings they say shyly, "How can we know?" Pressed, they ascribe them to the Old People or to Mungu (God), searching our faces in the hope of learning which one we prefer; our need to *understand* makes them uncomfortable. For people who must live from day to day, past and future have small relevance, and their grasp of it is fleeting; they live in the moment, a very precious gift that we have lost.

Lying back against these ancient rocks of Africa, I am content. The great stillness in these landscapes that once made me restless seeps into me day by day, and with it the unreasonable feeling that I have found what I was searching for without ever having discovered what it was. In the ash of the old hearth, ant lions have countersunk their traps and wait in the loose dust for their prey; far overhead a falcon—and today I do not really care whether it is a peregrine or lanner—sails out over the rim of rock and on across the valley. The day is beautiful, my belly full, and returning to the cave this afternoon will be returning home. For the first time, I am in Africa among Africans. We understand almost nothing of one another, yet we are sharing the same water flask, our fingers touching in the common bowl. At Halanogamai there is a spring, and at Darashagan are red rock paintings—that is all. . . .

The women are out gathering the silken green nut of the baobab, which, pounded on a stone and cooked a little, provides food for five months of the year. The still air of the hillside quakes with the pound of rock on rock, and in this place so distant from the world, the steady sound is an echo of the Stone Age. Sometimes the seeds are left inside the bell to make a baby's rattle, or a half shell may be kept to make a drinking cup. In the rains, the baobab gives shelter, and in drought, the water that it stores in its soft hollows, and always fiber thread and sometimes honey. Perhaps the greatest boabab were already full grown when man made the red rock painting at Darashagen. Today young baobab are killed by fires, set by the strangers who clear the country for their herds and gardens, and the tree where man was born is dying out. . . .

From a grove off in the western light, an arrow rises, piercing the sun poised on the dark massif of the Sipunga; the shaft glints, balances, and drops to earth. Soon the young hunters, returning homeward, come in single file between the trees, skins black against black silhouetted thorn. One has an *mbira,* and in wistful monotony, in hesitation step, the naked forms with their small bows pass one by one in a slow dance of childhood. The figures wind in and out among black thorn and tawny twilight grass and vanish once more as in a dream, like a band of the Old People, the small Gumba, who long ago went into hiding in the earth.

With the collapse of colonial governments, the destruction of wildlife by rampaging Africans had been widely predicted, and a glimpse of the last great companies of wild animals on Earth had been the main object of my trip to Africa in 1961. That year, Nairobi was still a frontier town where travelers to wilder parts were served and outfitted. Gazelle and zebra crossed the road at Embakasi Airport, and the Aathi Plain and the Ngong Hills, bringing Maasai Land to the very edges of the city, made it credible that the first six people to be buried in Nairobi's cemetery had been killed by lions.

Since then, the East African parks and game reserves have actually increased in size and numbers (though their future remains uncertain). Even the Congo's great Albert Park (now Kivu National Park), for which the worst had been foreseen, escaped serious damage. The one park destroyed by political unrest is the beautiful small park at Nimule, where civil rebellions, already begun when I passed through in 1961, have broken down all order. John Owen, a district commissioner in that region of the former Anglo-Egyptian Sudan who later became director of the Tanzania National Parks, flew over Nimule in a light plane in 1969. "A careful search," he told me, "produced nothing but one buffalo." Very likely the vanishing white rhino is gone forever from the Sudan.

Tsavo

We flew north over the waterhole at Mudanda Rock to the confluence of the Tsavo and Athi rivers, which together form the Galana, under the Yatta Plateau; from here, we followed the Yatta northward. This extraordinary formation, which comes south 180 miles from the region of Thika to a point east of Mtito Andel, is capped by a great tongue of lava; all the land surrounding has eroded away. The Yatta rises like a rampart from the rivers and dry plains, yet its steep sides present no problem to the elephants, which that day were present on the heights in numbers. The elephants of Tsavo are the most celebrated in East Africa, being very large and magisterial in color, due to their habit of dusting in red desert soil. Yet they were not always common here: the great ivory hunter Arthur Neumann, traveling on foot through the Tsavo region on the way from Mombasa to Lake Rudolf in the last year of the nineteenth century, saw no elephants at all.

Away from the rivers, the only large tree in this *nyika* [thornbush desert] is the great, strange baobab, but the baobab, which stores calcium in its bark, has been hammered hard by elephants, and few young trees remain in Tsavo Park. For many tribes, the baobab, being infested with such nocturnal creatures as owls, bats, bushbabies, and ghosts, is a house of spirits; the Kamba say that its weird "upside-down" appearance was its punishment for not growing where God wanted.

Lake Natron

For a long time I stood motionless on the white desert, numbed by these lowering horizons so oblivious of man, understanding at last the stillness of the lone animals that stand transfixed in the distances of Africa. Perhaps because I was alone, and therefore more conscious of my own insignificance under the sky, and aware, too, that the day was dying . . . I felt overwhelmed by the age and might of this old continent, and drained of strength: all seemed pointless in such emptiness; there was nowhere to go. I wanted to lie flat out on my back on this almighty mud, but instead I returned north into Kenya, pursued by the mutter of primordial birds. The flamingo sound, rising and falling with the darkness pinks of the gathering birds, was swelling again like an oncoming rush of motley winds—birds, bats, ancient flying things, insects.

The galumphing splosh of a pelican, gathering tilapia from the freshwater mouth of the Uaso Ngiro, was the first sound to rise above the wind of the flamingos. Next came a shrill whooping of the herdsmen, hurrying the last cattle across delta creeks to the bomas in the foothills of Shombole. A Maasai came running from the hills to meet me, bearing tidings of two dangerous lions—*"Simba! Simba mbili!"*—that haunted this vicinity. He asked nothing of me except caution, and as soon as his warning was delivered, ran back a mile or more in the near-darkness to the shelter of his *en-gang*.

I built a fire and broiled the fresh beef to keep it from going bad, and baked a potato in the coals, and fired tomatoes, and drank another beer, all the while keeping an eye out for bad lions. I also made tea and boiled two eggs for breakfast to dispense with fire-making in the dawn. As yet I had no energy to think about tomorrow, much less attempt makeshift repairs; the cool of first light would be time enough for that. Moving slowly so as not to stir the heat, I brushed my teeth and rigged my bedroll and climbed out on the car roof, staring away over Lake Natron. I was careful to be quiet: the night has ears, as the Maasai say.

Ngurumans

The Maasai Legaturi, disdaining the offer of a place in the two-man tent, had made himself a shelter out of thorn branches, but soon he came tugging at the tent fly, murmuring excuses, and once inside, he spat all over its triangular doorway of mosquito netting to bless this transparent stuff against the passage of night animals. Near [Philip] Leakey's camp, we had come upon a black-maned lion in the grass, and Legaturi, seated beside me, had hurled defiance at the king of beasts, splitting my ears with the blood-curdling whoops and chants used by the lion-killing moran of other days. The lion gazed at him, unmoved. When I drove closer, Legaturi subsided, grabbing up my binoculars and pressing them at me, imploring me to stop right there and take a picture: *"Simba! Simba mkubwa!"* (*Big* lion!) Closing the car window tight, he had shrunk into his blanket, glazed with fright. If Legaturi is a fair example, the agricultural Maasai have lost that aplomb with wild animals for which the tribe is well known.

In the middle of the night, a rhino blundered into us. A rude *Chough! Chough! Chough!* at the quaking canvas brought us both upright, and Legaturi seized my knee in a famished grip as if fearful that *l'Ojuju,* the Hairy One, might rush out to do single combat with the huge night presence whose horn was but a few feet from our faces. He did not let go until the rhino wheeled and crashed away. *"Kifaru!"* Legaturi whispered, finding his voice at last. *"Kifaru mkubwa!"*

Samburu

Winds of the southeast monsoon blew up from the hot *nyika,* and a haze of desert dust obscured the mountains. But the Uaso Nyiro flows all year, and along its green banks the seasons are the same. A dark lioness with a shining coat lay on a rise, intent on the place where game came down to water. At a shady bend, on sunlit sand bars, baboon and elephant consorted, and a small crocodile, gray-green and gleaming at the edge of the thick river, evoked a childhood dream of darkest Africa. Alone on the plain, waiting for his time to come full circle, stood an ancient elephant, tusks broken and worn, hairs fallen from his tail; over his monumental brow, poised for the insects started up by the great trunk, a lilac-breasted roller hung suspended, spinning turquoise lights in the dry air.

On a plateau that climbs in steps from the south bank of the river, three stone pools in a grove of doum palms form an oasis in the elephant-twisted thorn scrub and dry stone. Despite the wind, there was stillness in the air, expectancy: at the lower spring a pair of spurwing plover stood immobile, watching man grow older.

In the dusty flat west of the spring, ears alert, oryx and zebra waited. Perhaps one had been killed the night before, for jackals came and went in their hangdog way east of the springs and vultures sat like huge galls in the trees. With a shift in the wind, a cloud across the sun, the rush of fronds in the dry palms took on an imminence. Beyond the springs oryx were moving at a full run, kicking up dust as they streamed onto the upper plateau.

Climbing from the springs onto the plain, I crossed a stone ridge and stared about me. In every distance the plain was sparse and bare. Strange pale shimmers were far oryx and gazelle, and an eagle crossed the sky, and a giraffe walked by itself under the mountains. A Grevy's zebra stallion charged with a harsh barking, veered away, then circled me, unreconciled, for the next two miles, unable to place a man on foot in its long brain.

Northward, over pinnacles and desert buttes, the sky was clear, but directly ahead as I walked south, dark rain arose over Mount Kenya fifty miles away. Coming fast, the weather cast a storm light on the plain, illuminating the white shells of perished land snails, a lone white flower, the white skull and vertebrae of a killed oryx.

The sun, overtaken by the clouds, was sinking rapidly toward the Laikipia Plateau, and there were still four miles to go through country increasingly wooded; I hurried on. Awareness of animals brought with it an awareness of details—a shard of rose quartz, a candy-colored pierid butterfly, white with red trim, the gleam of a scarlet-chested sunbird in the black lace of an acacia. Set against the sun at dawn or evening, its hanging weaver nests like sun-scorched fruit, its myriad points etched on the sky, there is nothing so black in Africa as the thorn tree. . . .

A string of impala passed close at full speed, bouncing high; I hoped that no lion, having missed its kill, now sat disgruntled by the trail. At the edge of the woodland, fresh elephant spoor was everywhere, and inevitably there came the crack of a split tree that is often the first sign of elephant presence. None was in sight, however, and I hurried past. The red sun, in a narrow band of sky between clouds and mountains, had set fire to spider webs in the grass that while the sun was high had been

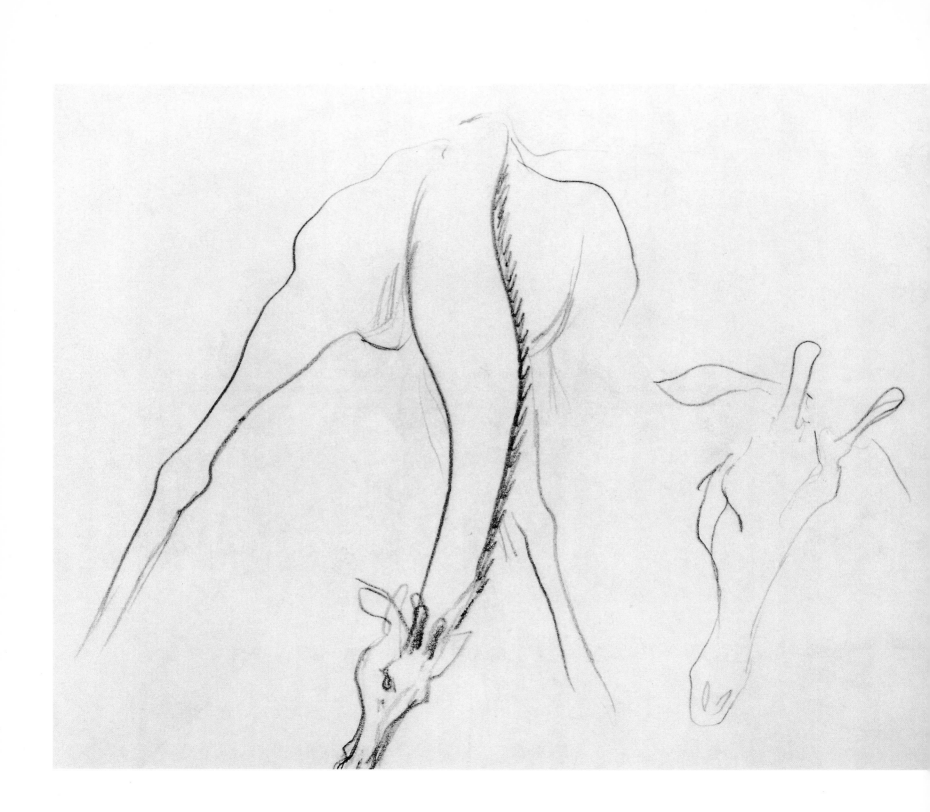

invisible; where I had come from, flights of sand grouse were sailing down to the Buffalo Springs for their evening water. Then the sun was gone, and across the world, a full moon rose to take its place.

The earth was still, in twilight shape and shadow. In the wake of the wind came the low hooting of a dove, and one solitary bell note of a boubou. I met no animals but the giraffe, a herd of eleven set about a glade, waiting for night. The giraffe were alert to my intrusion but in their polite way gave no sign they had been disturbed.

In camp, wild things were going on about their business—tiny red pepper ticks with bites that itch for days, and a small scorpion, stepping edgily, pincers extended, over the bark bits by the camp table, and ant lions (the larvae of the lacewing fly) with their countersunk traps like big rain pocks in the fire ash and sandy soil. Unable to find footing in these soft holes, the ant slides down into the crater where the buried ant lion awaits. A faint flurry is visible—the ant lion whisking sand from beneath the ant to hurry it along—and then the victim, seized by its hidden host, is dragged inexorably into the earth.

Mount Marsabit

This morning, the sun rising in the thorns looked silver and wintry in a haze and wind that made black rooks shift restlessly on the dead limbs. The silver sun was where the moon had been, in eerie light of day. The coarse bark of a gray zebra woke the plain, and the egrets went undulating southward, and oryx fled in all directions, cold dust blowing.

Mount Marsabit rises from the desert like a discolored cloud. Grassy foothills climb in steps toward isolated cones, and the air cools. In a meadow, like a lump from the volcanoes, stands a bull elephant with great lopsided tusks curved in upon each other, the ivory burnished bronze with age like a stone font worn smooth by human hands. This high oasis far from the old trade routes and new tourist tracks, and cut off in recent years by shifta raids, is the realm of the last company of great-tusked elephants in Africa. Many have tusks of 100 pounds or better on each side, and those of a bull known as Ahmed are estimated at 150 and 170 pounds, and may soon cross, as in the extinct mammoth.

Marsabit in June: great elephants and volcanoes, lark song and bright butterflies, and far below, pale desert wastes that vanish in the sands. On Marsabit are fields of flowers, nodding in the copper-colored grass: blue thistle, acantha, madder, morning glory, vetch and pea, and a magnificent insect-stimulating verbena, its flowers fashioned like blue butterflies, even to the long curling antennae. The blossoms of the different families are all of mountain blue, as if born of the same mountain minerals, mountain rain. One cow pea has a large curled blossom, and to each blossom comes a gold-banded black beetle that consumes the petals, and each beetle is attended by one or more black ants that seem to nip at its hind legs, as if to speed the produce of its thorax. Next day the flowering was over and the beetles gone.

An amethyst sunbird pierced my eye, and a butterfly breathed upon my arm; I smelled wild jasmine, heard the grass seeds fall. From the crater lake hundreds of feet below rose the pipe of coots, and the scattering slap of their runs across the surface. But no great elephant came down the animal trails on the crater side, only a buffalo that plodded from the crater woods at noon and subsided in a shower of white egrets into the shallows.

Sun and grass: in my shelter, the air was hot. Mosque swallows, swifts, a hawk, two vultures coursed the crater thermals, and from overhead came a small boom, like the sound of a stooping falcon. But the bird hurtling around the crater rim was a large long-tailed swift of a uniform dull brown. This bird, described as "extremely uncommon and local . . . a high-lands species that flies high, seen only when thunderstorms or clouds force them to fly lower than usual" is the scarce swift. Though not the first recorded at Marsabit, the sighting of a bird called the scarce swift gave me great pleasure.

Our camp was in the mountain forest, a true forest of great holy trees—the African olive, with its silver gray-green shimmering leaves and hoary twisted trunk—of wildflowers and shafts of light, cool shadows and deep humus smells, moss, ferns, glades, and the ring of unseen birds from the green clerestories. Lying back against one tree, staring up into another, I could watch the olive pigeon and the olive thrush share the black fruit for which neither bird is named; to a forest stream nearby came the paradise flycatcher, perhaps the most striking of all birds in East Africa.

Dida Galgalu Desert

Ahead, volcanic cones rose from the sand haze like peaks out of low clouds; the day was overcast with heavy heat. Larks and ground squirrels, camel flies and ticks; the camel fly is so flat and rubbery that it flies off after a hard slap. Occasional dry dongas support bunch grass and the nests of weavers; in this landscape, the red rump of the white-headed buffalo weaver is the only color. Though animals other than snakes are not a problem here, a lone traveler had made a small thorn shelter at the side of the road, to ward off the great emptiness. Round lava boulders, shined by manganese and iron oxides, and burnished by wind and sand, looked greased in the dry light—a country of dragons.

To the north, the Huri Mountains rose and fell away again into Ethiopia. We took a poor track westward. In the wall of an old riverbed was a cave of swifts and small brown bats where man had lived, and from the dust of the cave floor I dug an ancient digging stick with a hacked point.

Beyond Derati, gray zebra and oryx clattered across stone ridges, and a black-bellied bustard rose in courtship, collapsing its wings on the twilight sky like a great cinder in the wind. Then a striped hyena rose out of the rock, a spirit of the gaunt mountain: it turned its head to fix us with its eye before it withdrew into the shadows. This maned animal of the night, with its cadaverous flanks and hungry head, is the werewolf of legend come to life.

Lake Rudolf
[now Lake Turkana]

The head of a very large crocodile—*"Mkubwa sana mamba!"*—is spotted in the water off the point, a mile away. Through binoculars the surfaced snout and eyes can scarcely be made out; the fishermen cannot have seen the crocodile, only a rock that has no place along a shore that has been memorized for generations. The brute sinks slowly out of sight, to reappear some minutes later farther off; it raised the whole length of its long ridged tail clear of the water before sinking away again. Everywhere else, including the Omo River, the Nile crocodile is very dangerous to man, but here it seems to be quite inoffensive. Lake Rudolf crocodiles can leap clear out of the water, and have no difficulty catching fish; mostly they hunt at night, in the lake shallows. But they also have an excellent sense of smell, and will travel a long way overland for carrion. It is said that these ancient animals are unable to resist the call, *"im-im-im,"* of a weird nasality attained by closing off one nostril of the caller, but I found no chance to put this predilection to the test.

PART 2

TANZANIA 2

Summer 1976

In early July of 1976, under an old midmorning moon and a blue sky, we [Peter Matthiessen and Maria Eckhart, with her sister and two friends] set off from Njombe in the southern highlands on a safari of some sixteen hundred miles through Tanzania. We would go east to Chalinze, near Dar es Salaam on the Indian Ocean Coast, north to Mount Kilimanjaro on the Kenya border, west as far as the Rift Valley, and south again on the old dirt road that runs down the center of the country. This safari has no purpose, we are traveling just for the sake of traveling, but we shall visit the country's less-known parks—Mikumi, Kilimanjaro, Arusha, Tarangire.

Our vehicle is a safari Land Rover with long wheelbase lent by Dr. Gottlieb Eckhart, a sturdy veteran of the colonial days who has practiced medicine out here for more than forty years. Since I saw it first, in 1970, on a safari with the Eckharts to Ngorongoro and the Serengeti, its left side has been splayed like a sardine tin by a rhinoceros in Ruaha, giving it a certain raffish character. We have installed a plywood board there to fend off wind and thieves.

That first day we paused at Isimila, where on a lakeshore sixty thousand years ago Acheulian man manufactured his stone knives, scrapers, cleavers, and hand axes. Isimila today is a dry bed, like Olduvai Gorge, and the cores and flakes of much remote, dim labor lie by the thousands on its dusty floor; it is one of the finest Acheulian collections ever found. Now those early men are gone, the lake is dry, and little grows here; there are only faded butterflies, a parched east wind, the sad, relentless call of the ring-necked dove, a waiting in the air, as if Homo rhodesiensis, *slope-shouldered, suspicious, might raise his head among the boulders to watch us go.*

From Isimila the road passes down beneath Iringa, which sits high up on a plateau; it crosses the Little Ruaha River and winds down through portals of the Usagara hills toward the Kitonga Gorge. Here the road drops away from the southern highlands. Already the heat is rising from the valley, and now the first baobab tree appears as the road—having descended some four thousand feet—levels out in a new landscape of thatched villages. Although the day grows late, it is much warmer here, and the shine on the smooth, silver-lavender hide of the giant trees is soft. The vegetation of the plains appears—acacias and sausage trees; white-trunked Sterculia; doum palms; bright-green *mswaki*, called the toothbrush bush; bizarre pink-flowered desert rose. Bold yellow baboons shift along the road's edge in their sideways lope, and near Kidika a baboon troop has been run down. The baboons have been laid out in a line on the shoulder of the road, where the corpses are guarded from jackals and vultures by the lone survivor, a big dog baboon that glares balefully in its bewilderment as we rush by.

Beyond Kidika the road crosses the Great Ruaha, which continues eastward into the Selous Game Reserve; there it joins the Rufiji River on its way to the sea. The road follows the north bank of the river, where egrets doze on the pale, naked boulders of the dry season. Back in the trees shy elephants appear, for this region is one of the few places left in East Africa where large animals may be seen outside the parks, a harmonious land of blue, misty mountains, swift streams and green forests, and dry plains and giant baobabs scarcely damaged by elephants, whose numbers are in proportion to the habitat; because of malaria and tsetse people are few, and so this land looks nothing like the battered square miles in the parks, where the crowded elephants have taken shelter from human beings.

Now the road turns away toward Mikumi and the flood plains of the Mkata River to the north. In 1954, when this road was new, the Mkata Plain was laid open to the hunters and most of its trophy elephants and other large animals destroyed, but ten years later 450 square miles on both sides of the road were set aside as the Mikumi National Park. A sign reads HATARI! WANYAMA YA PORINI! ("Danger! Wild Game of the Bush!"). And we have scarcely crossed its borders when plains game begin to appear—impala, wildebeest, and buffalo, as well as zebras, elephants, giraffes, and a fine male lion. We make camp in Mikumi in late dusk, starting up a buffalo while collecting firewood; the possibility of such encounters is what lends character to a safari. The voice of lion and hyena come down in the night on the cool ocean winds from the southeast, and the chant of tree frogs, and the roar of trucks from the new road, like a hard welt on the old soft hide of Africa.

Mikumi is one of the smallest of the parks, and its animals are small as well, in particular the elephants; until 1964 it was a hunting tract, and perhaps the hunters shot out the big bulls with big ivory, leaving the smaller animals to perpetuate their genes. The Mkata Plain has a variety of terrains, from swamp pools and flat riverain savanna on the north side of the road to rolling hills of *miombo* to the south and east, and we saw Lichtenstein's hartebeest with its odd dark spot on the flank, and the black rhinoceros, growing scarce and wary almost everywhere, and such interesting birds as the painted snipe and martial eagle. But it is not a "wild park" like Ruaha or Tarangire, since its animals, confined to this small area by surrounding shambas, are as conditioned to visitors as those in Ngorongoro and Manyara, and seem set into these land-scapes as if placed on view. Perhaps Mikumi became a park too late; it seems little more than a small outpost of the immense Selous [pronounced Se-loo] Game Reserve. "The Selous," where I long to go one day, is generally regarded as the last great redoubt of wild animals left in Africa.

Ngurdoto Crater

While in Arusha I revisit Ngurdoto, the dead volcanic caldera between Meru and Kilimanjaro. Ngurdoto has not changed at all since 1970. The numerous buffalo and baboons, warthogs and water birds, scattered bushbuck, waterbuck, an elephant, a black rhinoceros with her calf, sedate on the water swale and meadow, might have occupied these very places on that afternoon six years ago when I walked across the crater floor with the askari Serekieli.

Because no one is present to forbid it, we go on foot around the east rim of the crater. On this bright morning, the forest of great olives and mahoganies overflows with sound and color—the squawk of touracos, the squalling of baboons, the amazing loud bark of the bushbuck, the puff and lowing of a hidden buffalo deep in a thicket near the road, the whistles and fluting of birds unseen, the rolling roar of a black-and-white colobus as it leaps up and down in primate rage on its leafy platforms. Perhaps this roaring male is so excited because of the white infant on the arm of its nearby mate; the mother, observing us from her green bower, remains calm, her thumbless hand on her infant's soft white head, her gray face in its striking cowl resigned. The splendid colobus is disappearing from the Earth because of the value of its pelt to its hairless kin; not only do the Chagga prize it for their traditional headdress in the way the Maasai prized the lion's mane, but it brings an ever-higher price in the revolting animal-parts bazaars of Arusha and Nairobi, with the result that this Ngurdoto population is one of the few that still remain. Perhaps threatened because we are on foot, the creatures crowd us (as baboons will crowd a leopard), and so we see three separate troops—a lucky morning. Bushbuck and waterbuck walk out upon the track only yards away, and on our return we slip quickly past a solitary buffalo—these are the dangerous ones—blowing in nervous irritation from its thicket.

Wild bovine dung and baboon stink, fresh minerals in forest humus, the frangipani fragrance of the *Tabernae montana*—a mix of smells that intermingle with the lavender sheen of the violet-backed starling, the brilliant blue and lemon of the variable sunbird performing a strange leaning courtship rite in which the perched male on its twig veers crazily around and upside down. In every shaft of light the insects spin—huge golden swallowtails and flame-colored Charaxes, papilios of white and yellow, gaudy bees of electric blue and red, orange and black. . . .

This hidden caldera in its virgin forest is one of the loveliest places in Africa, yet in a long day we see only one other vehicle, for this small park is little frequented, although it includes low woodland, the Momela lakes, and also the wild eastern slope of Mount Meru. Years ago I walked home along this road toward the Momela camp of my friend Desmond Vesey-Fitzgerald; at the end of a wonderful afternoon I found myself cut off by a herd of elephants and was happy indeed that Vesey came out hunting for me in the dusk.

West of the caldera, where the wood opens out into grassy hills, there is a wonderful view of the vast interior of Meru, laid open on this eastern flank by the explosion of what may have been the highest mountain in all Africa. Giraffes sway on the small hills, against white clouds. Around the shores and all across the open water of Big Momela Lake, a mile away, hundreds of thousands of flamingos fill the open water, necks crooked, heads upside down as they sieve the algae, in the weird electric mutter of these creatures, and long pink streamers cross the lake, forth and back and forth again in arabesques.

Soon the sun is firing the rim of Meru, and in one of those visions of East Africa that are scarcely believable even when seen, the streaming birds, rising into the last light, cross and cross again the snows of Kilimanjaro, which has risen from its wreath of clouds into the dimming blue of the twilight sky. How far away it seems; how very strange that just two days ago we were up there on the highest part of that snow mountain that looms over these tropics of the equator.

It is dark when we pass Vesey's camp; Vesey is dead. On the night road down the mountain we are stopped by silver-eyed giraffes, mute as great flowers; eventually they drift aside and watch us go.

Tarangire River

Tarangire, set aside in 1970, is the third-largest national park in Tanzania (after Serengeti and Ruaha), comprising about a thousand square miles in the valley and hills surrounding the Tarangire River. The river flows north from the Kondoa hills to die eventually in Lake Burundi, and in the dry season, from July through October, it attracts most of the wildlife in this vast, dry, open, rolling country that is part of the western plain of the Maasai Steppe. Like Ruaha, it is a "wild" park, with much tsetse infestation, limited access roads, and unsophisticated animals that cannot be counted on to show themselves.

At this dusty time of year, along the river animals are constantly in view; one might see almost all the resident large species once moving from this camp. Our cooking fire is perched just at the bluff edge, so that we may watch the river life below. Most of the herds come to the river after the morning grazing, but small bands of zebras, wildebeest, and waterbuck come and go all day. A pride of lions grunts now and then during the day and all night, too; in the dry season lions abide along the river. Therefore, the thirsty animals are skittish as they come and go, and the zebras yelp uneasily from the wood edge.

The big herds come again at night, filling the darkness with snorts and sudden runs. Somewhere between these river bends lions are waiting. With so much prey at hand, the lions make no secret of their presence, though most of their uproar

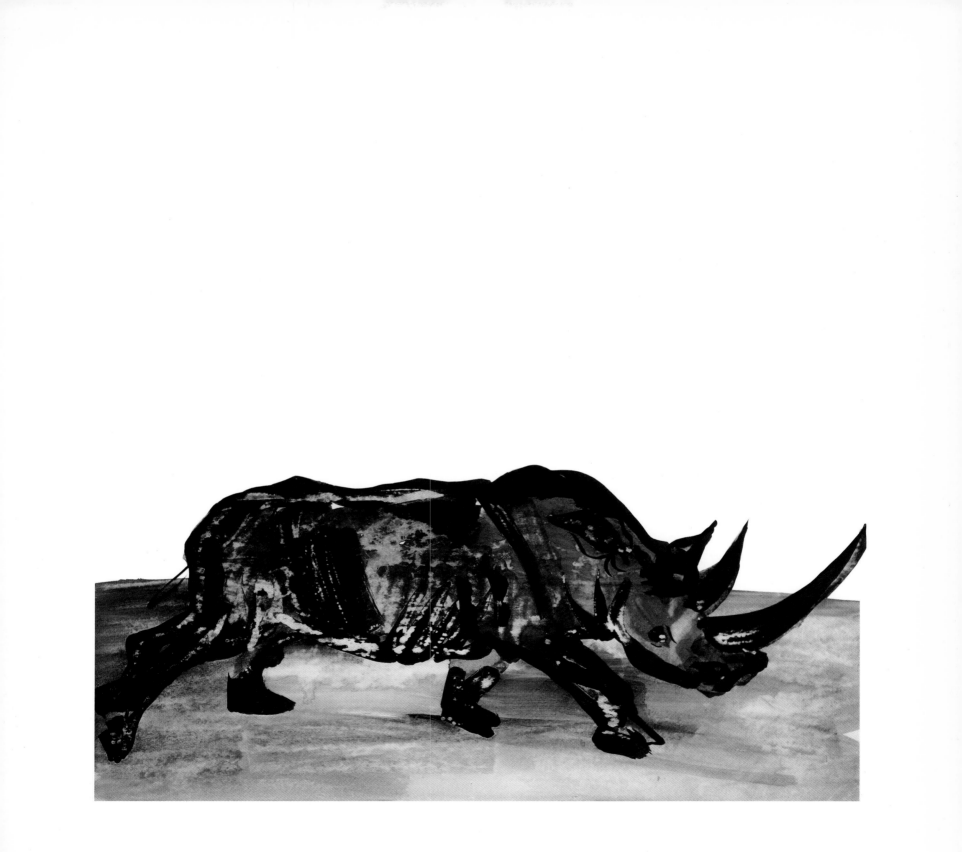

49

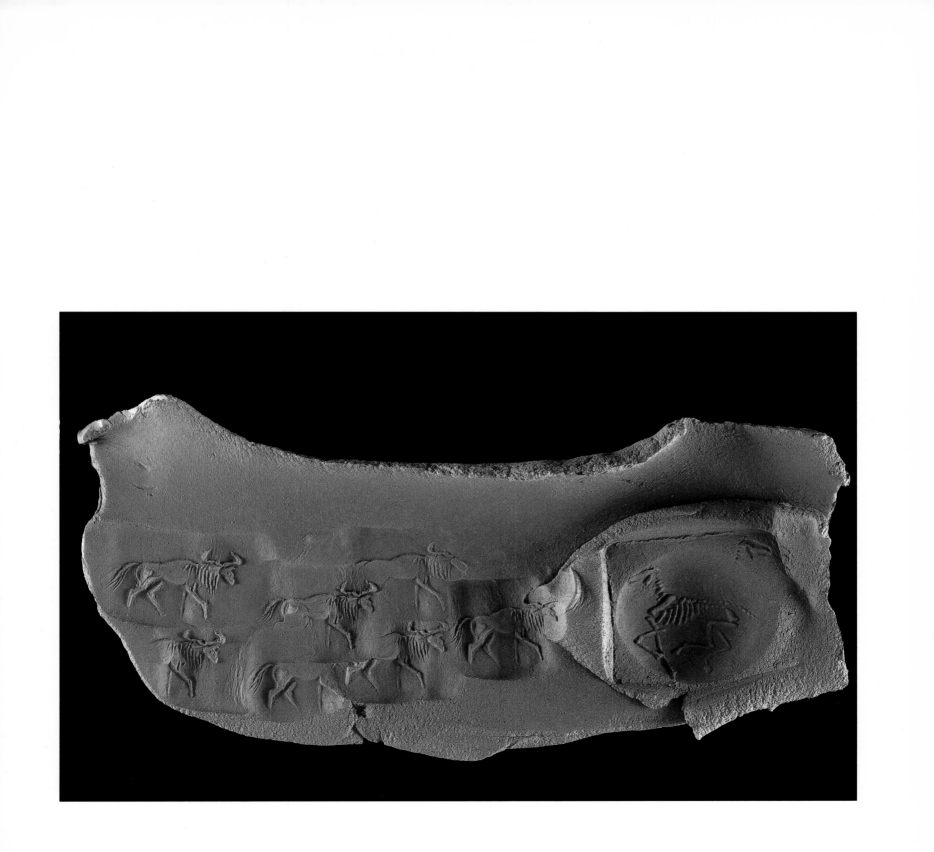

takes place after a kill. Hyenas announce themselves with that eerie whining call that sounds more like the cry of a huge bird, and one night they make their own kill near the camp, filling the still night with horrid giggling. The leopard comes, too, in its own time: the heavy grating cough that has been likened to the sound of sawing wood is heard each night, attended by the squalling of baboons. In the day both leopard and hyena remain behind.

The lions' ominous deep hollow din is interspersed occasionally with ostrich booming, for this is mating season, and the great legs of the adult males are turgid red. The cocks stalk along after the hens, which sometimes, in a resigned manner, squat low in acceptance but more often move indifferently away; then suddenly a hen is mounted, in a great ado and thrash of plumes and feathers that lasts a half minute or more, the cock's white-tipped undulating wings describing languorous S-curves in the dry air. His amours completed, he struts away without a backward glance, while the stunned hen may crouch awhile, breast flattened in the dust, as if to properly digest her rude experience.

The cock is unique among birds in the possession of a penis. At this time of year, the cock ostrich flushes red and tumescent in the neck and legs, and both sexes writhe and flounce and run. Careening about, they shuffle their fat wings on their backs like maids tying up apron strings while rushing to answer a bell.

In the acacias across the river is a small kopje that gives shelter to hyraxes and klipspringers. The hyrax looks like a sharp-nosed marmot, but on the basis of certain anatomical similarities, notably the feet, it has been determined that its nearest living kin are elephants. Perhaps as a defense against the attacks of eagles, it has the astonishing ability to stare straight upward into the equatorial sun. The rock hyrax is common and widespread, but the tree hyrax—grayer, with soft pelage—is less often seen, and I am very glad to find it here.

While young hyraxes of both species chase up and down the rocks and limbs, as if showing off for visitors, the klipspringers stand fixed on favored ledges only a few yards above the track, depending on the stone gray of their coats, the tawny underparts—so close in tone to the sere light that falls through the dry leaves of commiphora—to keep them hidden. Spike horns, harlequin ears, and bright black eyes dissolve in the rock and shadow. But sometimes one seems to acknowledge being seen by pettishly stamping its short legs, which are designed for swift, sure springs from rock to rock.

The smaller mammals, so interesting and so often overlooked, are much in evidence at Tarangire. Besides the hyraxes, bush squirrels and ground squirrels are both common, and so are three species of mongoose—the banded, the slender (black-tipped), and the pygmy, a tiny bright-eyed rufous form that finds protective coloration in the termite mounds. There is also a glimpse of the large marsh mongoose, or possibly the clawless otter, looping through the rank riverain grass under the camp. One night a bold genet cat comes to our fire, and bats of unknown species are quite common. The black-backed jackal is abundant (one gave ground reluctantly when I surprised it digging up some buried meat), and one afternoon on the eastern plains three bat-eared foxes look us over before loping away into the whistling thorn.

At this season the Tarangire River is no more than a series of shallow pools connected by the disappearing stream, although one pool is a near-lake of sedge and water lily. There reedbuck and waterbuck feed submerged up to their shoulders; when their heads are down, the reedbuck look like mounds of warm brown earth among the lily pads.

Dead buffalo and elephants, hollowed out by jackals, are otherwise virtually intact; one elephant lies like a huge mummy by the river. Not far away are two more skeletons, stripped of their ivory; we take a clean skull back to camp as stool and table. That so many died so close together—and also the extreme wariness of all the rest—means that ivory poachers are at work in Tarangire.

At midmorning two hundred buffalo rush in dusty avalanche down to the river east of the camp, scattering the zebras and wildebeest that were there ahead of them and subsiding with loud puffing and blowing in the shallow water. Zebra and wildebeest return, and other creatures come quickly from the woods, until the low pools splash—warthogs and baboons and waterbuck, six eland, a solitary oryx, two pairs of ostriches, and finally a small band of kongoni. But suddenly the nervous zebras yelp and gallop, scattering the rest; the shy kongoni never reach the stream at all.

The lion pride lies in thorn shadow where this bluff descends into the river; they pay no attention to the nervous animals. With binoculars, standing a few yards from our breakfast fire, I can count fourteen. All have fat bellies, and most are sound asleep, in piles, big paws dangled in the air. Most are lionesses with grown cubs with faint spots and a dark young male.

It is near noon now, the heat settles, and sweat flies nag at the damp face, but the tsetse that have plagued us in the woodland are absent from this open country with its cool wind from the southeast. Overhead, pale griffons and dark bateleurs turn and turn again, and below, three reedbuck lie in the rank grass, seemingly untroubled by the lions. The day is cloudless, the sky pale with wandered dust that rises in small swirls on the southern horizon and casts its pall on the far, dark Rift escarpments to the west. On a hillside of killed trees a line of elephants move in somber company through the pale grasses.

In the stream a waterbuck noses along after its cow, and another cow with a half-grown calf approaches the river opposite the torpid lions. In midafternoon, two lions in vain pursuit of waterbuck rush across the water, then lie down on the riverbank and gaze about them. Everywhere stand herds of animals, stiff-legged, ears high; then, with danger past, they whirl, and rush away, and the dust rises from the hot woods of afternoon. Next morning at daybreak the dead tree nearest the lions is laden with still vulture silhouettes, and twisted and torn remains lie near the track. Whether this young waterbuck was the same animal that had been rushed the day before we do not know.

Not long after dark a herd of buffalo comes to the river, just under the camp. The night before, many wildebeest had moved into the shallows, making a constant swishing sound, like tropical rain; tonight, their nervous blaring at the edge of the great buffalo herd is the only sound. Even the wildebeest fall still as in the hot night, full of dread, the buffalo move in utter silence in the river. The huge wild cattle, so cumbrous in their ways, as if counting on weight and horns and numbers to deter attack, are suddenly so strangely quiet that tree frogs and locusts' shrill are heard over their sound, and the hum of the campfire, and a warm wind in the black leaves, under a crescent moon. In awe we stare down upon the dark forms just beneath; not one beast puffs, not one steps heavily into the water, all seem to tiptoe.

August 1977

Ruaha

The track comes down off the escarpment into the hot thornbush wilderness in the center of the country; through this near-desert the Ruaha winds like a long oasis. It is late August now, the dry season's bitter end, and in mysterious anticipation of the rains the dead gray spiky bush, set off here and there by the white, leafless skeleton of a sterculia, is coming into flower; it seems magical that any root could find moisture enough to send color and form through these dry stalks.

Toward noon the track winds down around a hill, and the Ruaha comes into view, brown, slow, and shallow, between green walls of great trees that go coiling down across the country to the Rufiji River and the sea. The river forms the southern boundary of the park, which in its five thousand square miles is second in size only to the Serengeti, but only a small area here in the southeast has road access or tracks of any kind. Formerly the park was part of the mighty Rungwa Game

Reserve, but the rough country, with its tsetse and malaria, was discouraging even to hunters, who made small impact on the elephant armies. Yet elephants learned quickly that they were safer inside the park than out, and became so concentrated around the only permanent water—the Ruaha itself—that at the time of my first visit here, in early 1969, only five years after the park had been established, Ruaha was already being mentioned in the great East African elephant controversy that had started at Murchison Falls National Park in Uganda in 1965 and spread to Kenya's Tsavo National Park. There David Sheldrick, the warden of Tsavo East, having saved his elephants from Waliangulu poachers, was trying to save them from the scientists, who were trying, in turn, to save them from themselves.

We make camp on an open bluff, under the great stately acacias known as winterthorn. The rust-colored curled pods have begun to fall, a great attraction to the elephants, and I am some distance from the camp, looking for a way down the steep river bluff to fetch water, when the first one comes, upwind of me and moving fast; to a man on foot, the dimensions of an approaching elephant are quite astonishing. It does not stop when I wave my arms, and I run away along the bluff, having no confidence in its intentions. But the elephant gains, so I stop again and leap and shout, not wishing to jump over the bluff, and this time it sees me, or gets my scent, and performs a graceful, silent change of course, with no show of alarm or threat, gliding away inland from the river.

In 1969, in winter, Douglas-Hamilton and I swam in the swift Ruaha torrent, but in summer the stream is hot and flat, sliding slowly across the continent between the broad sandbars of the dry season. On the bars are the striking water birds of Africa, and most of the noisiest are here—the fish eagle and Egyptian goose, the hadada, the wattled plover, and the water dikkop, which saves its noise for night. Though speckled pigeons come and go, land birds are few: here as elsewhere along the river, the elephants have destroyed the bush, aside from the bright-green toothbrush and the ancient winterthorns and figs and tamarinds, too huge to be dismantled or pushed over. Nor is there any grass along the river, no place at all where a lion might lie up: the river is wide open to the game, which is somewhere in our view throughout the day, and the impala are attended by young fawns, sometimes nine together. A huge troop of banded mongoose lives behind the camp and moves along the bluff, searching out food.

The Mdonya flows only in the rains; in the dry season it becomes a pale sand river, winding down between steep flood-cut banks to the Ruaha. Except in drought there is water not far below its surface, and the wells dug by elephants in these sand riverbeds are later resorted to by other animals, but because of the absence of surface water, animals are few, and the Mdonya bush is relatively undamaged. Farther to the west, in tsetse woods, there is a fallen elephant, oddly similar in its collapsed aspect to one of the fallen baobabs that its own kin have hollowed out and killed—a great sprawl of exhausted matter, of organic waste. Probably this animal was a drought victim, for these remains are a few years old. Someone has made off with the ivory, and the feet are separated from the carcass, as if the scavenger had hacked them off to sell to tourists as cute wastebaskets, then given up this rotten idea in the filth and heat.

Dispirited by flies and heat, we return to camp. In the late afternoon a huge, black sky arises in the south, toward the Msembe Mountains, and in stark light a solitary giraffe comes down to water, peering around for a long time before splaying itself in the only pose in which a grown giraffe is vulnerable to lions. Even so, we are astonished when a lioness bounds from the shadow of the tamarinds, and the giraffe, alerted by an impala's snort and a wild clamor of zebras and baboons, struggles to rise. There is too much bare ground between giraffe and lioness, and she quickly gives up this desperate foray, returning slowly to lie down again in open shade under the tamarind. We are sorry for her, for the hunting here is hard. She is not sorry for us, however; she gazes across the river at our camp. Not far away a second lioness is crouched against a tree, watching us, too, and we wonder if these are the two we saw last evening, padding along the river bluff at dusk. Next day, on our own side of the river, one makes an unsuccessful stalk of warthog, passing close by three sullen buffalo that lift their heads to lower at the lioness but do not run. The second lioness is somewhere near, for from the heavy dusk by the river comes uproar and the drumming flight of the impala; a herd of waterbuck runs out onto the plain, and baboons shriek.

Ordinarily, park lions are well fed, but a lioness that will make a run at a full-grown giraffe is probably working too hard for a living, due to the lack of cover for an ambush. The lions here are deadly silent—also a bad sign—and so we pay more attention than usual to the cluster of *mswaki* bush just behind camp.

From the river's edge come two large herds of cow elephants and calves, as a third herd comes in to water from the south. This a mighty spectacle, since several hundred elephants are now in sight, yet it is disheartening, as well. The destruction of their habitat has not yet slowed the elephants' reproduction, to judge from the great number of young elephants in view. Hurrying along, the infants flap their ears like wings, as if this might help them to keep up as the two herds move rapidly away. But those on the far side of the river, having not yet scented us, go on about their business, as if one herd paid no attention to another; we come up quietly under a winterthorn and get out of the Land Rover without disturbing them, although we are not sixty yards away. They dust and blow and water, vast guts rumbling, and some rub their baggy rumps on riverbanks and trees—on one old stump the surface fairly shines. Then a great bull, a little downwind from the herd, gets our scent and goes off blaring; perhaps he is only an old nuisance, for the cows pay no more attention to the bull than they do to us.

Then the wind shifts, the trunks go up like question marks to identify the scent, and the whole herd wheels away in dusty flight, silent at first, then clamorous; so alarmed are the great beasts that they break ranks and scatter. Part of the herd soon reappears, a mile upriver, still trumpeting in consternation; the elephants cross the Ruaha just above its confluence with the Mangusi, and the heavy wash of their wild passage through the water comes downriver like a rush of storm.

On a river bend, against the Msembe Mountains, a giraffe stands quietly in the Ruaha as if transfixed by its own reflection. The sight is strange. This is the first time I have seen a giraffe standing still in open water, and we wonder why this experiment is taking place. . . .

Once again, this afternoon, black thunderheads assemble on the mountains to the south, though the season is too early for the rains, and once again the storm light appears haunted. We return upriver. On the far bank, a lioness dead still, stands emblazoned against the fierce green of an *mswaki;* had it not been for the green, we would not have seen her, for she is the color of dry ground. Four more lionesses lie partly hidden in the bush, and all are facing down the river, where a herd of greater kudu comes slowly down to drink. The kudu have chosen a dangerous place, just under a steep bank, and they move carefully; a magnificent bull stands guard above, together with a calf, while the seven or eight cows jump down to water, nor does he descend until most of them regain the bluff. Even then, he turns his head before he drinks, light dancing from the lyrate horns as he peers about. A skimmer lilts, oblivious, up and down river, as the lionesses crouch low in their bush. They scent the kudu, but they are too far, there is no cover, and they make no run; restricted to these few places of ambush, they can only watch the animals that circle wide around the thickets.

And now there fall a few drops of rain, big heavy drops that gather up balls of gray dust in the black cotton soil and obliterate the oblong holes of scorpions. Dust settles, the air freshens, the world is cast in drastic light. The sun descends. The lions wait.

At dark come sudden wind and dust swirls that drive us to take shelter behind the tents. It is impossible to keep a fire going, and so we make do with cold food and strong drink. In the absence of fire the animals are bold or curious or both, and after dark the elephants and buffalo draw near. Taken aback by the pale tents, they stand there a few yards away like huge, black shadows in the moonlight that floods through airy canopies of thorn. For a long time they are completely still; we remain still, too. A hooded vulture on a nest that blots the stars shifts on its white sticks as it peers down; a scops owl gives its gentle trill as if unable to bear the suspense of such great silence. Then the elephant that frequents this grove—the one that sent me flying that first day—comes up under the bluff, just beneath the tents, round feet crunching softly on the sand. The bluff is too steep for him to climb, so I stand my ground a few feet above his head; we regard each other, sharing unknown dimensions of the night.

The wind has died with the water dikkop's cry, the stars are clear. When the animals go, I lie down in our tent. At night in Africa one sleeps well and also lightly, for among the familiar noises of the night there is always one that seems entirely strange. From downriver comes the deep, hollow grunt of a giant eagle owl, the nasal plaint of the hadada, the restless ringing whistle of a greenshank from Eurasia lifting its wings for the night journey farther south. And now the sound comes, a liquid, falling song, never heard before and too far away to follow, in this night country stalked by hungry lions. The unknown sound is trilling, and it makes me sad.

Deep into the night I am awakened by the roar of lions, then the caterwauling of hyena—the familiar ululating whoop of a spotted hyena from over the river, and from nearby a bark and moan that surrounds a series of horrified, muffled cries, like calls for help from a woman being strangled. Spotted hyenas snigger on a kill, but this other sound is entirely strange, and it makes me sit straight up on my cot. When at daybreak I walk out, the plain is empty; the plains game is scattered and there is no sign of a kill, only the fresh pugmarks of lion and hyena among the round and wrinkled disks made by the elephant.

In all their visits to the Ruaha the Eckharts have never laid eyes on a hyena, which elsewhere is not shy and is often seen; furthermore, they camped here many times before they ever came across a lion. Eck attributes the scarcity to heavy lion shooting in the days of the Rungwa Game Reserve, but perhaps the destruction of the cover by the elephants has more to do with the steady increase of lion sightings; on this safari we see lions once or twice each day. This morning there is a pair of lions near the camp, a male and a female, but they have no dry blood on the muzzle, nor pendant belly, nor other sign of having made a kill, and they are moving when we find them, at a time of day when well-fed lions are lying up in shade. In the disconcerting way of lions they do not look at us or increase their pace; it is as if man were not there.

In a squall of birds a fierce sun rises from the crown of a great winterthorn across the river. Beyond, acacia savanna stretches south and east to the Msembe Mountains and the escarpment of the Southern Highlands, open country and a land of eagles—harrier eagles, coursing in quick pairs along the wood edge, and the long-crested hawk eagle, solitary, dark, in shadowed ambush on a favored limb, and the great martial eagle, on the very crown of a giant tree on the eastern sky.

Here one is nourished by each breath of so much space and light, such hard, clean air. There are no sounds here but wild sounds; in three days we have not seen or heard another vehicle. Dung smell, fresh heat, the brown glitter of the river; in a light wind I hear the fall of an acacia pod, bearing seeds that will be devoured before a tree can grow. A slate-colored boubou skulks about the toothbrush bush, and a few barbets, woodpeckers, and ashy starlings come and go. "This is the way it was," Eck sighs, staring away over the empty country. He is seventy this year; next year he will leave Africa for good. He has made many safaris to the Serengeti and to Ngorongoro, but the Ruaha is his favorite place in Africa.

At midmorning in this cold winter weather, our light airplane [Peter Matthiessen and Victor Emanuel are guiding a wildlife safari to Botswana] rises out of Gaborone, Botswana's capital, which lies across the frontier from South Africa. Soon the last imprints of man—sand tracks, boreholes, the faint outlines of abandoned farms—die out in the gray thornbush that occupies almost all of Botswana except this narrow edge in the southeast, as well as large contiguous regions in Zambia, Angola, and Namibia. This is the Kalahari (Kgalagadi, or "mother of thirst"), an arid steppe some three thousand feet above the sea; to the west lies a true desert, the Namib, that extends to the dunes of the Atlantic Coast. The seasonal changes of temperature in this channel determine the climate far inland, and the Kalahari has a season of light rains from November to May. Thus there is growth—a hostile thornbush, very like the nyika of East Africa—but the soil, overlying aeolian sands that in places are four hundred feet in depth, is thin and fragile. Here and there one sees bare, gray depressions, and in the rainy season there is water in these pans; besides water, the pan floors supply a vital residue of the trace elements that the herds require. But now, toward the end of the dry season, the pans are dry, and animals have moved north and east toward the Chobe drainage and west toward the Okavngo swamp, for there is no permanent surface water anywhere in Kalahari.

The airplane's shadow drifts across the central desert, where pans are few and the thornbush much diminished, thousands of square miles without track or scar or any sign that man has been here—one of the few land areas left on Earth of which this can be said. Even the Bushmen, who leave no mark, come rarely to this central desert . . .

Okavango River

The bush thickens and deepens, green coils of riverain forest rise from the flat landscape, the eye is relieved by sparkling black water. Here in the southern part of the swamp the Okavango is a labyrinth of creeks and islands, marsh hummocks and broad savannas of the flood plain; but as the airplane moves north, the channels gather, and the current is seen to run more swiftly between green walls of papyrus, grassy swales, and white beaches in the big bends of the streams. The swamp spreads off to the east, and the extent of it cannot be seen, for the air is shrouded by bush fires that over the years have made a patchwork of the Okavango.

Eventually the channels gather into main streams that join and part and join again; quite suddenly the Okavango narrows and becomes a river, like the neck of a flat flagon, although it is still so far across that even from the air the eastern shore is only dimly seen. Where the thorn scrub of the Kalahari draws close to the west bank, Yei villages give way to Mbukushu. Ahead lies a dark blot of river forest. In the forest lives the great orange owl that takes live fish from the river.

Late on the first day at Shakawe a big orange shape is seen, crossing deep shadows in the bower of the huge dark elephant's-foot trees that dominate the forest: since many ornithologists have sought the fish owl without much success, we are delighted and yet unfulfilled—the glimpse has been too brief. In the bright winter air and river light and wind all around the forest, we marvel at huge star-shaped puffballs and the sweet song of a robin chat, three squirrel heads protruding from one hole, a python between the riverbank and the high reeds. In the dry season insects have shriveled, and there is no humidity, but one must keep an eye out for black mambas.

One night we went by boat upriver, hoping to spotlight the fish owl in a river tree. In a short swale at the river edge, the beam picked up a pair of hadada ibis, eerie bronze-green, red-eyed, mute, on a dead limb, close enough to touch, and a big gray gymnogene, bare pink-orange face peering out from its odd bonnet of erect feathers, made no effort to escape, remaining motionless on the fat limb of a fig. Then the swinging light caught a pale blur and swung back again; for a few precious seconds, there it was, a very big round-headed owl pinned like a huge orange moth against the blackness of the forest wall. Unlike the other birds, it was not blinded, and it vanished into a black hole in the trees.

We sit a while in the silent dark, under southern stars soothed by night winds in the reeds along the river. A hippo bellows in the distance, and there comes a deep wave of memory, almost an echo, of the first time I ever heard that sound, fifteen years ago on the White Nile, in Equatoria. And indeed, Ngamiland is very like that part of the Sudan, where a vast and hostile thorn scrub and savanna lies on all sides of the green reeds that enclose the clear dark water of a river. Water lilies, wind and light, the kingfishers and swallows, the huge water birds, the broad circle of raptors in the sky—all are the same. And here, too, the drums resound, the soft thudding on the skin of Africa, in the liquid dark over Ruare, which hisses through cool deserts of the night.

Chobe

The Gubatsa Hills of the great Chobe National Park in northeastern Botswana overlook the Savuti River—formerly a course of the Chobe River, now a channel that carries the flood overflow of the Chobe into the great Savuti marsh. The clear green flow of the Savuti winds south into its marsh, and beyond the marsh shine the hot game plains of Mababe. I try to imprint upon my memory this mighty prospect of hundreds of square miles of ancient Africa, this view all around a vast horizon without the smallest sign of human life. There are only blue mountains, water, trees and dust and sun, the eagles and winter elephants and morning antelope whose patient movements emphasize the space and silence. There is no time here, yet all seems to wait, like the silent paintings of the vanished hunters in the rocks below.

It is mid-August now and deep into the dry season of late winter. Through thin shade of *morula* and paperbark and the graceful *Terminalia sericea* that Selous called the "silver tree," one can see the giraffe and elephant that the small hunters saw, and from the hilltop kudu and buffalo are also visible. On the far bank, broad wet back gleaming in the morning sun, a hippopotamus stands motionless, resting its heavy head upon the sand as if lost in thought.

I scan mile upon mile for sign of sable, which look black at any distance and would jump out at the eye from this faded landscape that spreads away north to the brown hills; on the far side of the hills is the Chobe River, already close to its confluence with the Zambezi, which it joins not far above Victoria Falls. To the south lies a broad flood plain, the Savuti Marsh, and beyond the marsh lies the Mababe Depression, a famous hunting region in Selous's time, when the great game concentrations of South Africa were already gone. Flying in toward this region from the west, we saw a cavalcade of buffalo that raised dust from the open woods for well over a square mile.

Game is not yet as plentiful on the Savuti as it will be just before the rains, when the river is reduced to scattered pools, but in camp there is rarely a moment of the day when animals of some sort are not in view. Buffalo and waterbuck are common, and a herd of greater kudu comes and goes away again, almost unnoticed. The impala, warthog, and giraffe that are so plentiful elsewhere in the Chobe do not seem to like it in these woods, although the thickets have been opened out by elephant and there is little cover for a lion. Vervets and dark chacma baboons are almost constantly in sight; across the stream noisy dog baboons lead their troop back and forth under the trees, and this year's infants jump off their mothers' back to play or scamper along big fallen limbs, squeal, tumble, run, and nag to be slung up again behind; even these pink young carry the tail in the "broken" way so characteristic of their family. The infants are observed by the pair of African harrier eagles that hunt this wood, moving swiftly through the trees like huge accipiters. The old bull elephants of the plain come to the wood in search of the big velvety pale pods of camel thorn; a bull wedges a tree trunk between his tusks, as high as possible, and shakes the tree until the pods rain to the ground. At the south end of the camp, just beyond our tent, is a main thoroughfare where elephants and buffalo come down to water. The watering elephants crop the fresh green grass along the banks, dark clay sand that is excellent for dusting is available on the woodland slope just a few yards away, and there is an ideal crotch between two rough limbs of an acacia where an enterprising elephant may scratch its ears by inserting its head, lowering it a little, and withdrawing.

A male lion, then another, cross an open grass sward at the very edge of high marsh swale. Both are strong and well-fed creatures with glowing hides, and both are wary; one moves off into the marsh as we approach, while the second crouches so close to the earth that its gold-chestnut mane, fretted with black, flows away among the blowing grass tips. As the Land Rover draws near, the lion opens its mouth wide in warning, then rises stiffly and follows its companion, picking up big cat feet with distaste as it enters the water. Probably these two are wanderers, unattached to any pride—the lions that belong here are not wary. Later in the long grass country not far north of the Mababe Depression, we watch three males with a group of ten lionesses and young animals, none of which bother to stand when we appear: the only ones that move at all are the two pairs that are mating. The males snarl and paw at the females, walk side by side with them around the vehicles, sit restlessly, and stand again as the females snarl at them in return; lion mating may take place intermittently for hours, even days, and these two females, perhaps unsettled by the stink of vehicles, are getting cranky. But both submit, mating occurs, and the whole round of restless courtship starts again. Taking advantage of the situation, the remaining male, a bit long in the tooth, approaches a third lioness, then moves off with a jaded expression when one of the younger males turns to look.

We move southward. Savuti is a country of great raptors, and in the drowned trees by the marsh nine bateleurs sit in silent company; the sepulchral young, a dirty gray, may be identified from afar by their heavy cowls. Under a large acacia, not far away, lies a dead chacma baboon; although it has been there for some time, nothing has touched it. Under rare circumstances—is it disease?—vultures and hyenas avoid a carcass that ordinarily would attract them, for even these undiscriminating creatures have their limits. Perhaps the baboon was killed, dragged high into the tree, then dropped there by a leopard—it has that broken air about it.

Like most crane species of the world, the wattled crane of southern Africa is now uncommon, and we are happy to locate five of these white birds whose English name is the only inelegant thing about them. As they step carefully along the marsh edge, the wattle is scarcely apparent; what one remembers is the long sweep of the neck, the horizontal back that is common to all cranes, accentuated in this species by a dark line that separates the bone white of neck and breast from the pale gray-blue of the wings and mantle. Days later I see the five again, stately and beautiful, consorting with impala in the sun and shadow of lawnlike glade under an open canopy of acacias. . . .

Toward noon the sun appears, and at this time of day when in other seasons the animals lie up in the bush, the creatures move around in the warmed air. Where we had passed before and seen nothing at all, a large cow-calf herd of winter elephants

browses calmly in low, open woodland, as if it had been there since time began. Unlike the wild herd farther west, these elephants pay us no attention; perhaps those others had come in from the dangerous hunting tracts outside the Chobe. A white flurry of helmet shrikes in a silver tree, a flow of pale movement close at hand as three cheetahs bound away among the bushes, and startled giraffes break into their rocking run across a hillside. Around the next bend are sassaby, wildebeest, impala. In just a few minutes the empty land is filled with animals magically summoned up by the warm sun.

In silver deadwood of big elephant-killed trees, in a dramatic silver light under silver clouds, a long line of dark animals awaits us. They are sable antelope, twenty-five or more. We come up slowly, and the striking creatures do not bolt; they only move gradually into low bushes, where they stop and turn again and look us over. The leader is a mighty bull that looks shining black from head to tail except for its white face marks and white belly; deeply curved, with heavy base, its horns seem twice the size of others in the herd. The animals move off a little, turn again, and the dark heads with their recurved scimitar horns line up in patterns. They gaze at us, we gaze at them, nostalgic for that peaceable kingdom before time began.

At a riverbed shaded by winterthorn, a gust of small birds blows into a thicket. These wandering parties of small birds of many species are common in Africa except in breeding season—suddenly a dead-quiet place will come alive. Besides the fork-tailed drongo, which is often leader, this one includes tits, crombecs, camaropteras and eremomelas, some weavers in cryptic winter plumage, a prinia, a puff-back flycatcher, and the striking sulfur-breasted bush shrike. Not far away, a band of helmet shrikes is harrying a pearl-spotted owlet, not with any serious intent, it seems, since they soon lose interest and fly off. The rumpled owlet composes itself, dismissing man's gaze with an infinitesimal shrug.

Owls are various and common in Botswana—at least, we have seen many on this trip. Besides the spectacular fishing owl at Shakawe and Xugana, there was an African marsh owl near Lobatse in the south, the tiny barred owlet at Shakawe, the ubiquitous scops owls and pearl-spotted owlets, and the great eagle owl (the Verreaux's eagle owl of East Africa) observed here at Savuti. At dusk last evening this huge congener of the North American horned owl was silhouetted on the sunset sky in a dead tree; the day before we had seen one blowing on a windy perch at the edge of Savuti marsh. Today I met one not five feet from the ground, in an acacia thicket; the sinking sun gleamed through its ear tufts as it sat in silhouette in its cave of thorn, gazing down upon a flock of helmeted guinea fowl that wandered about, oblivious, beneath it. Perhaps the owl had waited untold owl time for these dim-witted birds to make this sad mistake, only to have a Land Rover jerk to a halt near-by, back up, and swing around, scattering the guineas in the process. The owl itself seemed philosophical and did not move; observing the intruders in a series of slow winks, it bared the eerie, milk-lavender eyelids that cause it to be called *Bubo lacteus*.

A pair of eagle owls frequent the wood across the stream from camp, and like so many African birds they perform duets, in this case an exchange of slow, lugubrious grunts like giant frogs, not song by our own definition, but no doubt music to their ears. This morning the owls have caught the eye of helmet shrikes and a shrill pair of brown parrots, which join forces to harry the thick, heavy birds through the morning trees.

In a glade two pairs of impala bucks are sparring. When one buck is defeated and runs off, the victor charges at the second pair; neither of these is a serious contender, both run off, too. At this the triumphant buck gives a wonderful display, a series of exhilarated stiff-legged bounds like the "pronking" of the true gazelles, as if jumping up and down in rites of spring, but the does that he has won pay no attention.

Earlier this month, in Kenya's Maasai Mara, I had seen a migrant greenshank south from Europe. Yesterday I saw one at Savuti marsh, and today there are many more, in company with other Palearctic migrants, such as the wood and the marsh sandpipers and the little stint. In this season the first Eurasian birds down from the north encounter migrants from the south; this sacred ibis by the pond edge nests in the high veld of South Africa and has come north to escape the austral winter.

On the stony summit of Gubatsu I perch a little while in the warm sun. This fly that comes to rest upon my arm is not a tsetse, which in this dry season has scarcely bothered us at all; one day "fly" may disappear entirely. That day may be the beginning of the end for large African wildlife, unless the rampaging unrest in all these countries ends it first, for tsetse alone is responsible for the existence of the Chobe park, for the Kruger in South Africa—the last refuge of large animals in that vast country—and the Luangwa in Zambia, and for almost all the wildlife parks and game reserves throughout East Africa; and fly is the last defense of these wild regions against the hordes of domestic stock that scour the continent.

One tends to forget that parks and reserves also protect forest, water, and undamaged land, which in these febrile political climates, surrounded by famine-haunted man, will go under quickly once the tsetse is defeated. Human needs, so it is said, must be placed before those of wild animals—though in the long view the distinction is a false one—but to encourage human increase and consumption when human numbers are the scourge of every continent is not to help humanity but to do it harm, a point that would-be benefactors making progress against the tsetse fly might keep in mind. One day, if balance is restored between resources and populations, and if the black man, questioning the white man's dangerous progress, rediscovers his kinship with this ancient land, a modest sanctuary should be set aside for the rubbery flat flies of the genus *Glossina*, in the absence of which African wildlife would be all but gone.

At Cape Verde, at the westernmost point of Africa, in Senegal, the ocean sunrise, clear red-blue, turns an ominous yellow, and the sun itself is shrouded, ghostly, in this dusty light of the northeast trade wind of the dry season, known as the harmattan, *that blows across the great Sahara desert. We depart in the heat of afternoon, proceeding east through the red earth wastes of Dakar's suburbs with their fringes of thin eucalyptus, introduced all over Africa to replace cleared forests and combat the vast erosion that threatens to blow the whole continent away. This region between desert and savanna, called the Sahel, is a parched thornbush of baobab and scrub acacia, red termite hills, starlings, doves, and hornbills, very similar in character to the* nyika *of East Africa, but as the road moves south and east, this thornbush rapidly gives way to open woodland and long-grass savanna, known to ecologists as the Sudanian zone, that separates the Sahel and the equatorial forest for four thousand miles across the continent to the Nile basin. Because terrain and climate are so consistent, the composition of its fauna is consistent, too. Our [Peter Matthiessen and primatologist Dr. Gilbert Boese] journey would be a preliminary inquiry into what remains of wildlife in West Africa, and our first destination, some three hundred miles inland, was Niokolo-Koba, in the southeast corner of the country near the borders with Mali and Guinea-Bissau.*

SENEGAL AND CÔTE D'IVOIRE

March 1978

Niokolo-Koba was West Africa's first national park at the time of its establishment in 1925, when after centuries of remorseless slaughter, the wild animals were all but gone. By 1920, the last damalisk (the western topi) in Senegal had disappeared, and both giraffe and elephant were near extinction. A solitary elephant killed in 1917 was the last one in the Cape Verde region, and since there are no recent records for Mauritania, the population in Niokolo-Koba represents the last elephants in northwest Africa. With protection, they showed good signs of recovery, and their numbers may now exceed three hundred, but in recent years, the plague of poaching that has done such damage to East Africa has set in here, with special attention to elephant and crocodile, and so despite the efforts of two-hundred-odd askaris who patrol on foot, bicycle, and by pirogue, the elephants are declining once again.

Though a few scattered animals still live outside of it, Niokolo-Koba—which has doubled in size since 1925, and presently includes 800,000 hectares—is the last stronghold of large animals in Senegal. The relict creatures of the region were spared by the remote location in the southeast corner of the country, in unsettled tsetse woods well south of the main trade routes. Watered by the upper reaches of the Gambia River, the park includes Sudanian savanna as well as the high tropical forest typical of Guinea, which lies just over its south border, and therefore maintains the only population of wild chimpanzees in Senegal. Besides elephant and chimpanzee, Niokolo-Koba can claim several hundred Derby eland, largest of all African antelopes, as well as the statuesque roan antelope for which the park is named: Niokolo-Koba means "Place of the Roan Antelope." (The roan is not easily seen now in East Africa, where it seems to be declining due to epizootic disease.) But, in the 1960s, the giraffe became extinct, and an effort to reintroduce them from Nigeria came to naught when a cargo of ground nut waste intended to feed the captive group was sent back by mistake, so that the creatures perished in their cages. Otherwise, the large mammalian fauna is—or was—quite typical of the Sudanian region, all across West Africa, including buffalo, hippopotamus, warthog and bushpig, and such antelope as the western kob, the large western hartebeest, Defassa waterbuck, bushbuck or "harnessed antelope," Bohor reedbuck, oribi, and a few species of duiker. For many of its species, if not most, Niokolo-Koba can claim the most northerly as well as westerly populations on the African continent. Officially, at least, all the large predators are here (although the status of the cheetah is obscure), and the smaller mammals include five primate species; besides the chimpanzee and the baboon, there is the patas monkey or "red hussar," the green vervet, and the lesser gallego, or "bushbaby."

In the outlying areas of the park, where they are most accessible to poachers, animals are few; instead, one sees the round clay cylinders of the dead villages whose lands were appropriated as the park enlarged its boundaries. Then troops of *Papio* appear—the thickset, reddish nominate race called the "Guinea baboon." Because European scientists came here early, many of the original descriptions of African fauna and flora derive from Senegal; hence the prevalence of the specific name *Senegalensis* for such widespread creatures as the Senegal cuckoo, which I first saw in Botswana, thousands of miles to the southeast, and renewed acquaintance with this morning at the dump behind the Estekebe Hotel in Tambacounda. Since most of the early naturalists were French, the hartebeest is *le bubal*, the buffalo *le buffle*, and the kob is called *kob de Bouffon* (after the eminent taxonomist of the eighteenth century).

A white-tailed mongoose, a patas monkey, vervets, then the red-flanked duiker, stamping black feet and flicking its tail straight up and down as it regards us: like all duikers, it has short horns and short forelegs and holds its head low to the ground—adaptive characters for quick escape in the dense bush or forest that duikers prefer.

The slow river of the dry season is clear and green, setting off huge rocks exposed along the bank and reflecting the soaring fan palms or borassus; this high dark gallery forest by the river is a riverain extension of the Guinea forest to the south. Here in the heat of the midafternoon the elephants come down to water, and hippos may be seen not far upriver. The shy, small forest *buffles* have been there, too, to judge from the bovine dung along the way.

In the dead heat that persists into the dusk, the kob and waterbuck lie down on the dry mud of Sita N'di (the western kob seems to ignore the hottest sun) but the bushbuck and warthog have retreated into the dry shade of the woodland all around. Here and there, the woodland floor is white with silk-cotton from the ceiba pods, which are eaten by baboons as well as vervets and thereby scattered in the time of seeding. Over the white woods hangs a ghostly stillness, set off by hot wafts of the *harmattan* in the dry fans of rapphia and borassus. Bamboo the browning color of burning white paper has somehow sprouted from a crust of lateritic stone, and the common *Pterocarpus* tree sets forth a pretty yellow blossom, as if in anticipation of the rains.

In a grove of huge figs, by a dark creek of stagnant water, a company of beasts has gathered—for the shade, perhaps, for we never saw one drink. A big roan buck leads a band of hartebeest out of the woods to join a rabble of baboons and vervets, a pair of bushbuck, and a pair of oribi the color of brown grass. In East Africa the oribi are reddish, and thus these creatures seem to be an exception to a general tendency toward erythrism that characterizes a number of West African forms: at Niokolo-Koba, for example, the bushpig, bushbuck, buffalo, and baboon are all markedly more red than their counterparts in East and South Africa. The colobus monkey of West Africa is also red, and so is the pygmy hippopotamus of the river forests of Liberia and Côte d'Ivoire. Why this should be so is quite mysterious; the conspicuous color would seem to be an evolutionary disadvantage. But in early times, the forests were much more extensive than they are today, and perhaps most or all of these "red" animals once inhabited the forest, where animal colors tend to be brighter than on the savanna. (Why *this* should be so is also a good question: perhaps it has something to do with social communication and/or epigamic display in a dim light.)

Farther south another roan crosses the track, but I am distracted by a long-tailed parakeet, spectacular bright emerald in its color; as with many birds of the savanna, its range extends across the continent into northern Kenya. The parakeet flickers rapidly through the dry air, alighting at last among white flowers of a vernonia bush at the edge of marshes. Not far away, by the west sump of a dry pan, an extraordinary conference of birds has gathered, as if reconciled by drought to their great differences—speckled pigeons, laughing and vinaceous doves, the red-billed wood dove, black magpies and gray hornbills, the long-tailed and the purple glossy starlings, cattle and squacco egret, and, on stalks of reed, the Abyssinian and blue-bellied rollers. The last species excepted, all of these birds or their close congeners may be found somewhere in East Africa; it is the makeup of the group that seems extraordinary.

Near the Guinea frontier, a track turns off the sea and enters the Parc National de Casamance, a coastal rain forest dominated by figs and palms. Gratefully we leave the car and walk about on foot. Though the day is warm, the sea forest remains cool, its deep shade thinly filtered by the sun. We find the print of a small bushbuck or a large yellow-backed duiker, hear the telltale puff of what might be a nervous *buffle* back in the forest, but here as at Niokolo-Koba, this beast eludes us. The only mammals seen, in fact, are squirrels and monkeys—green vervets and the guenon, or mona monkey, that handsome red and black relation of the Central and East African blue monkeys. The rare western red colobus remains hidden—this is the monkey I most wish to see. The paths are strewn with tamba, the small brown monkey-apple, which is relished in these parts by every anthropid, from these small *Cercopithecuses* to *Homo sapiens.*

Where the forest subsides into red mangrove estuaries behind the coast, palm-nut vultures have convened in the most seaward of the trees—striking white birds that have mostly abandoned the vulturine habits of their kin and subsist largely on nuts of the palm oil, in the vicinity of which they are usually found. I was astonished to see one alight on mud along the estuary and waddle about among the mangrove stilts in pursuit of fiddle crabs and perhaps mud skippers, both of which abound on the tidal rivers. No doubt this is a well-known habit of this species, but I shall record it here in case it is not.

Early next day we crossed the Casamance River by small ferry and drove north into Gambia on a rough road across the airy coastal plain. We continue on foot along a tongue of gallery forest that follows the dry bed of a stream across the open fields to a shaded place of boulders and damp sand—a rainy-season pool where animals still come to dig for water—and have scarcely arrived there when a large troop of western red colobus that have been observing us burst forth in reckless aerial display from the high treetops, scolding and barking, the long-limbed silhouettes hurling themselves from bounding limbs or sailing downward in wild arcs and careening into the dry bushes as if intent on tearing down the forest. The forest clearing on all sides has confined them to this narrow tongue of trees, but I am delighted that my first sight of this striking creature—it is black above, rich chestnut-red below—should occur in the African countryside. Even in East Africa, there are few places anymore where one may see such animals outside the parks.

The red colobus here at Saleite will not survive long. Less than a mile farther on, the people are burning down the forest, and a huge high crackling flame, riding the wind, roars through a copse of high trees near the road. The fire is attended by kestrels, Abyssinian rollers, and cattle egrets: the egrets stalk about in the flame's path, intent upon the spearing of small fugitives, while the rollers and pale orange falcons hover and dart like departing spirits in the smoke, the harsh racket of the rollers a fit accompaniment to the violent crackling of the blaze. . . .

On the flight to Côte d'Ivoire, the carry-on baggage of one Senegalese lady consisted of three large and springy fish. The tails of these whoppers refused to fold down neatly, and kept flipping up the wings of their cardboard carton. For lunch we were served "bush meat"—more precisely, some small cold birds with gloomy sizzled heads smeared over with what one could only hope was pâté. Otherwise the flight southeast over Guinea forests was uneventful until, circling wide over the sea on the approach to Abidjan, there came into view the reddish sands and long unbroken line of heavy surf that spared this Windward Coast (Liberia and Côte d'Ivoire, which lay windward of the slave ports of the Gold Coast—modern Ghana) from the worst depredations of the slave ships.

Abidjan is a modern city that on this fetid, humid coast retains all the dirt, smells, and decrepitude of the old slave ports. We were eager to leave Abidjan for our destinations at Parc de la Comoé, in the savanna, and arranged for berths on the evening train for Ougadougou, in Haute-Volta, which would let us off just after daybreak at Ferkessedougou, in northern Côte d'Ivoire. From there, in a small red auto, we set out on the fifty-mile trip east over rough dirt roads to the Parc de la Comoé, a tract of more than 2.8 million acres set aside in 1953 as the Bouna Game Reserve and made a national park in 1968, eight years after Independence.

We had been warned about the lions of Comoé, one of which had mauled a careless client. A lepidopterist whom we met later in the Parc Marahoue, to the south, claimed to have seen a lion and a leopard at Comoé, as well as the obligatory *buffles* and *bubals* (or perhaps it was *kob de Bouffon* and *bubals,* for here, too, the crafty *buffles* eluded us), and no doubt we failed to give proper attention to this park, which, excepting the chimpanzee and giant eland, is said to contain most of the large mammals that are found at Niokolo-Koba. Yet by standards of that park (which would be thought meager in East Africa or in Botswana), Comoé is thinly populated indeed, not only by large mammals but by birds. In three hours, in late afternoon, a dozen kob and *bubals* showed themselves, also an oribi, a warthog, and three distant hippos in the Comoé River, which flows all the way south to Abidjan. Of elephants or buffalo, or even their manure, there was no sign.

In view of the alarming decline of animals throughout Côte d'Ivoire, a commendable law against shooting any kind was passed in 1974, but this law is ignored by the Lobi hunters whose villages surround this northern part of the park. The Lobi, who have successfully resisted Islam and Christianity alike, bring their families into the park during the rainy season when the tracks are too muddy to be patrolled; they build their square huts, plant yams and millet, and hunt very much as they have always done. Originally a wild people from Haute-Volta, they take it very much amiss when their old ways are interfered with, and only a few years ago, the warden at Comoé felt so threatened that he prophesied to several friends—correctly, alas—that these Lobi would take his life. Understandably, the present warden has made no better progress against poaching, if these lonely remnants of the wild populations that could be supported in this huge, well-watered park are any sign.

Not that the warden would have given us much hope about the prospects for the Parc de la Comoé, or of any park in all West Africa. Zoologists assume that the wild ungulates of western Africa were always less common than in the east and

south, not only in numbers but in species. The rhinoceros, if it ever occurred here, vanished long ago, as did the wildebeest and zebra; for reasons not well understood, West Africa lacks the astonishing variety of antelope that is found south and east of the Nile. Edaphic poverty is sometimes blamed, but the weakness of tropical lateritic soils is fairly consistent almost everywhere throughout this fragile continent, including the game plains in the east that support the greatest biomass of animals on Earth, and perhaps the imbalance is better explained by a more simple reason, one that suffices easily all by itself: south of the deserts, in land inhabitable by man and beast, West Africa has far more human beings than East, and human beings have been here a great deal longer, hunting and trapping, burning and cultivating, competing for the pasturage and water, and eroding and exhausting the poor soils. As early as 1934, in a book on his travels in West Africa, one observer remarked, "But I should have been surprised if I had been told I would travel about seven thousand miles without seeing any live wild animal larger than an antelope." There are many fewer animals in West Africa today than there were fifty years ago, when the white man's tools and weapons became widespread, but the decline had begun many centuries before.

This open grassland of small, scattered trees that resist annual drought as well as fire is all but monotypic from Senegal east into Central Africa; the fauna (and flora) are essentially the same, not only in Niokolo-Koba and the Parc de la Comoé but also in the Mole Reserve in northern Ghana; in a complementary cluster of wild parks—l'Arly, Pendjari, and "Parc de W"—near the common borders of Haute-Volta, Niger, and Benin; and in parks in northern Cameroon. Since most of this region has been human's domain for thousands of years—and the rest will become so as tsetse is brought under control—

there isn't much hope that the status of wild animals in any of the West African parks would differ much from the status here in Comoé, and the few reports would indicate that it does not, despite the establishment of "reserves" and even some token restocking programs (as in Nigeria). Unlike Kenya and Tanzania, where the ambivalent attitude toward wildlife is quite similar but where an economic impetus based on tourism is clear, these countries see no good reason to protect what is left of their wildlife, far less to restore what is now gone.

Except in regard to Senegal, good data on the status of wildlife in West Africa is rather scant—a reflection in itself of official attitudes—but a comprehensive survey made in Nigeria in 1962 confirms most of one's worst fears about this region. Excepting the vast and empty lands of southern Sahara—Mauritania, Mali, and the Niger—Nigeria is much the largest of the West African states, and because it adjoins the states of Central Africa, such as Cameroon, where human beings and their weapons are less common, and where a reservoir of wildlife still exists, it might be expected to be better off than the smaller countries to the west. But according to this report, the last black rhinos in Nigeria were exterminated in either 1935 or 1945—no one really knows, which says a lot—the giant eland and several other antelopes have vanished, and almost all of the remaining larger mammals and even the jackals, are threatened with extinction, together with the larger reptiles and large birds. Of the thirty-two hoofed species, all but nine are extinct, threatened with immediate extinction, or "seriously depleted"; the exceptions are the bushbuck and a few species of duiker. Except in its only game reserve, Yankari, a tract of about eight hundred square miles northwest of the Benoue River where the fauna is now said to be increasing, it is unusual for the visitor to Nigeria to see any live wild animals.

In East Africa, the loss of habitat through intensive settlement, land use, and overgrazing has been the main threat to wildlife; but in countries such as Nigeria, which tolerates year-round hunting (often at night and often in gangs) of every species, regardless of scarcity, sex, or age, together with epidemic use of steel traps, snares, and encircling fires, the outright destruction of the animals themselves may be more damaging. Out of seventeen animals in a collection made in recent years by a Monsieur Brandt, sixteen had been previously wounded by crude pellets from one of the estimated four million or more muzzle-loaders used in the backcountry for hunting "bush meat." In this populous, poor countryside, wild game has always formed a high percentage of the meager protein diet; thus, *nama* the local Hausa word for "animal," also means "meat" (apparently this word survived the southeastern migrations of the Bantu-speakers out of Nigeria and Cameroon, since *nyama* is the Swahili word for "game"). This makes good sense to most Nigerians, as was noted by that traveler in 1934: "One is continually being reminded by Nigerians that theirs is the most densely populated nation in Africa and that perhaps, therefore, there is no place for wildlife." Although Nigeria has passed commendable wildlife legislation, it has not bothered with the education of the public, which has an understandable contempt for laws that are not enforced.

West of Ouazamon, the red road enters a new land broken by inselbergs of round black boulders, the granite outcroppings of Africa's old mantle. We pass an ancient hunter with his muzzle-loader, an old woman of Niger selling medicines, a solitary patas monkey near the road. The hundreds of miles of rough dirt track that we would travel in this land was not once crossed by a baboon; nor did we see the Abyssinian ground hornbill that was so common in backcountry Senegal, where roadside villages are at least as numerous and other wildlife as scarce. This turkey-size hornbill is venerated here by the Senoufou as a primordial animal and is a common subject of their carvings, but even this privileged status has not spared it.

Odienne is a Malinke stronghold in the northwest corner of the country, near the Mali border, a high and open town with a white mosque, set in low hills. Of the native woodland, there is little left. At noon the dust is bright, and the hot

wind of the *harmattan* blows unimpeded through the naked branches of the flame trees. From Odienne a track goes north to Bamako, on the Niger, but the main road turns south among citrus groves, guava, and cashew. From hilly terrains of thickening vegetation flow thick streams, as the savanna gives way to tropic forests. In the distance, tall pale boles of teak appear at the edge of the forest wall, and at the forest edge are birds, but no mammal is seen or heard, only a band of twenty hunters armed with ancient guns, marching empty-handed home along the road. In the absence of public education, local hunters tend to believe that wildlife is inexhaustible until it disappears entirely, and rarely make the connection between wildlife and habitat that might keep them from using fire as an aid to hunting.

The Ivoiriens assure us that unbroken forest is all that may be seen at Marahoue, but a large part of this 250,000-acre park is comprised of grassland and small hills set about with woods. In the deep forest, we observe magnificent butterflies of several genera and a white-collared mangabey *(Cercocebus)*, an angular gray monkey with a reddish crown, sitting sprawl-legged in a tree crotch on the farther side of a green meadow by the forest wall, keeping company with its smaller cercopithecine relatives, the mona monkeys. Atop a hill overlooking open country, we came upon a lepidopterist, caught in an illicit act of lepidoptery, who testified nervously that hartebeest and buffalo had been in view only this morning, but though we poked about the tracks until late afternoon, we saw no sign of *buffle* or *bubal*, nor even one memento of the elephants that are supposed to be here, too. The best that can be said for all too many of these "forest reserves" in the West African countries is that one cannot absolutely deny the presence of large animals any more than one can prove a negative proposition; how can one say they are not there, since one cannot see them?

In the great humidity, under heavy gray skies, we felt depressed. It seemed to strike us all at once that in hundreds of miles of travel overland in Côte d'Ivoire, the only animal we ever saw outside the parks was that lone monkey on the westward road from Boundiali, that the greatest concentration of wild mammals we had seen in the whole country were the fruit bats in the city park in Abidjan. Nor did anything that we could learn of other countries in West Africa promise much better.

Earlier today, passing the road that leads toward the park at Grand Lahou, Jacob Adjemon, our interpreter, had said, "In Abidjan, when we wish to regard elephants, we simply fly in an airplane to Grand Lahou, and there look down upon them." The driver, Mamadou, at first impressed but now fed up with the grand airs of his compatriot, had stared at me to see if I believed this arrant nonsense. But perhaps Jacob had sensed that we were saddened by the disappearance of wild animals from Côte d'Ivoire, as well as by the many signs that for all the "reserves" that have been set aside, for all the governments' proclamations of intention, the fatal destruction of West African wildlife still continues. The arrays of steel gin traps in the Man market, the gangs of hunters on the roads, the "bush meat" offered in backcountry restaurants, the unobstructed poaching that, for lack of serious intervention, will soon destroy the remnant creatures and thereby aggravate the serious protein lack in all these overpopulated countries—this obliteration of the native fauna is a crucial loss throughout West Africa, for reasons that go deeper than the "conservation of a priceless heritage" that white well-wishers like to prate about, having practiced it too late in their own lands. The animals are the traditional totems and protectors of the clans, the messengers of the One God that most Ivoiriens still perceive in all creation, the links with the world of the unseen, with the cosmic balance. Now the animals are gone, or at least so scarce that they have no reality in daily life. And perhaps even an urban boy like Jacob Adjemon, who has not bothered to go home to his Bete people in the last eight years, who is proud of the huge Hotel Ivoire and disdainful of "the bush," now grows uneasy. And so he says in a bored voice, "In Abidjan, when we wish to regard elephants, we simply fly in an airplane to Grand Lahou, and there look down upon them."

The Virunga National Park is named for the volcanoes that lie to the south of the Ruwenzoris—the Mountains of the Moon—just below the equator between East Africa and northern Zaire. As early as 1889, Leopold II had set up preserves to save the elephants from black people in order that they might be killed by whites, but the Virunga Park—the former Albert National Park, in what was then the Belgian Congo—was the first true national park in all of Africa; it was established in 1925 on the recommendation of Carl Akeley, who in 1921 had collected five gorillas here for the American Museum of Natural History. Since then the park has been considerably enlarged, before and after independence was declared in 1960. When I first came to Africa in the early winter of 1961, it was assumed that the Albert Park and all its animals were being ravaged and destroyed by the hordes of insensate Africans who were making life so miserable for the colonials, but this report turned out to be as exaggerated as many others, thanks largely to the park's African guards, who went unpaid for several years and defended what is now Virunga against the worst of the depredations.

On Saturday morning we arranged a ride to Nyiragongo, the southernmost of the Virunga volcanoes. Several of the "fire mountains" are still active, and only a year ago, on January 10, 1977, Nyiragongo quite suddenly erupted. As five coulees, or lava rivers, poured down its steep sides, the entire Bahutu village of Bukuma utterly vanished and more than two hundred people died. A few scorched skeletons of trees still stand in the shining fields, and these are being chopped for fuel by survivors of the cataclysm, who straighten here and there, in silhouette, to watch us pass. There is no smoke from the volcanoes, and on islets of high ground new gardens and banana groves have been established; but one day in the not far distant future, Nyiragongo—the mother of the spirit Gongo—will erupt again.

Climbing the hill, we look across to dense plantations in Rwanda. This region of volcanic ash forms a rich and well-drained soil—one of the few good soils in all of Africa—and Kivu Province, despite civil war and economic setbacks, continues to produce good crops of bananas, tea and coffee, cinchona, and pyrethrum.

Augustus Gabula, the young Bahutu warden who guided us uphill on Nyiragongo, was near the summit when the 1977 eruption took place and fled down through a tongue of forest between lava rivers, his path broken by an elephant herd that was stampeding off the mountain. Perhaps because, as mammalogist Jean Dorst informs us, "Having the legs straight with the bones placed vertically one above another, they are quite incapable of leaping," not all of Nyiragongo's elephants survived. Augustus led us two or three miles up the major flow to a place where a group of six beasts had been overwhelmed—not overtaken but asphyxiated as the lava flow burned up the oxygen on the mountainside. Among the hollows in the lava left by burned hardwood trees are scattered a group of elephant molds containing large white bones. In one of the graves the whole pachyderm form is still discernible, even the holes made by its tusks (long since removed) and a sad, curved tube of stone where the trunk lay. These were forest elephants (once considered a distinct species, due to the smaller size and small rounded ear), and some of them were young. Although there is no forage near the graves, only coarse bracken, marigold, and thorny acantha, elephants still make their way out onto the cooled lava and commune for a time with the six encased in stone, to judge from the copious amount of dung around the grave sites.

Toward noon, clouds shift and rain comes blowing through the forest, leaving behind a hot and humid sun. The Kivu-Ruwenzori chain is the heart of the African highland-forest habitat, which has outposts to the east in Ethiopia, in the Kenya highlands, and on mounts Meru and Kilimanjaro in northern Tanzania, as well as far to the west on Mount Cameroon. In all of these places many species of the flora and fauna are the same. Within the forest the tree limbs are thickened by moist gardens, mostly fern and orchid, and flowers of many kinds occur—yellow composites, pink impatiens, peas, the gloriosa lily, and a large hibiscus with blossoms of a dark lavender. Strangely, butterflies are few, and other than elephant trails, with their fresh dung, there is little sign of animals. But a small troop of the beautiful L'Hoest's monkey *(Cercopithecus l'hoest)* barks at our appearance, then retreats with dignity across the tongue of lava escorting a female with young clasped to her belly. This semiterrestrial species is a shining black with a bright-chestnut oval patch from the shoulders to the base of the long tail and fluffy white whisker. Compared with its relatives, the mona and blue monkeys, which are widespread in West and East Africa, respectively, the L'Hoest's monkey has an odd, small, scattered range, being confined to the mountains of eastern Zaire, Mount Cameroon, and the island of Fernando Poo, in the Gulf of Benin.

Kahuzi-Biega

The mountain gorilla sanctuary called Kahuzi-Biega is twenty miles above Bukavu at the south end of Lake Kivu, and at seven in the morning a taxi was found with gasoline enough to take us up there, though not back; the driver planned to coast down all the way.

Since arriving in Bukavu three days ago, we have not seen a single tourist and had hoped to have the gorillas to ourselves. But as luck would have it, two carloads of Belgian visitors turn up right behind us at the village of Bashi Bantu people that marks the entrance to the park. We have permission to camp here overnight, and the head warden, or *conservateur assistant,* has promised us our own guide for tomorrow, but today only one guide is available, and so all visitors must stay together. Disgruntled, we walk through the Bashi village and follow a path cross-country on the mountain flank to the farthest point where the others may be driven, for these six people who are to be *nos copains de safari* intend to take with them a very large brown-and-yellow-plastic ice chest full of lunch. Rather than lug it up the mountain by its handles, the unfortunate African assigned to this box steps into the bushes and with his panga cuts some strips of flexible green bark for "bush rope"; with this he rigs himself a tumpline in order to carry the big chest on his back.

Though it is still early, the day is hot and humid. Our little band, following the three small Batwa trackers—*les pisteurs*—pushes through tangles of coarse bracken, elephant grass, lianas, and cane between the forest and the overgrown plantations; since there is very little forage in "pure" forest, the gorillas are drawn to the abandoned fields of the Bantu peoples' shifting cultivation, where the sun encourages a variety for their own good. The trackers descend into swampy streams and up again into the forest, investigating the paths made by the apes and the freshness of their droppings; since gorillas are entirely vegetarian and must eat vast amounts by way of fuel, the droppings are abundant, large, and rather greenish, with a mild sweet smell. In midmorning there comes a sound of cracking limbs from a tree copse on the far side of a gully; the small *pisteurs* are pointing with their pangas. But one of our *gaie bande,* a silver-haired man who looked flushed even before he started, has not kept up; he is back there doubled up over a log, suffering heart flutters. Meanwhile, I am warning his compatriots about nettles, about the sharp spear points made by panga cuts on saplings, about false steps, mud slides, safari ants—

"Ngaji!"

The first gorilla is a large dark shape high in a tree, in a mass of stillness that imagines itself unseen. Then, near the ground, a wild black face leans back into the sunlight to peer at us from behind a heavy trunk, and the sun lights the brown gloss of its nape. Soon a female with a young juvenile is seen, then—*le gros male! Voila!* There is a big excitement as a huge silver-backed gorilla, rolling his shoulders, moves off on his knuckles into the tangle. The shadows close again, the trees are still. In the silence we hear stomach rumbles, a baboon-like bark, a branch breaking, and now and then a soft, strange "tappeting" as a gorilla slaps its chest; this chest-slapping is habitual and is only rarely used by way of threat. . . .

A Klaas's cuckoo sings, long-crested haws eagles in courtship flight sail overhead, and from the thicket comes the sweetish chicken-dung aroma of gorillas, accompanied by low coughs and a little barking. It is a standoff. On one side of this big thicket, perhaps thirty large and hairy primates are warning the restless young among them to be quiet, and on the other, a like number of large, hairless ones are doing the same thing. But all at once the suspense is broken by the ceremonial opening of the plastic ice chest, which incites a rush upon the lunch; the meat sandwiches and hard-boiled eggs that appease the hairless carnivores assail the platyrrhine nostrils of the hairy herbivores back in the bush, for there comes a wave of agitation from the pongid ranks. The thicket twitches, shifting shadows and a black hand are seen; the humans stop chewing and cock their heads, but there is no sound. The gorilla, like the elephant, is only noisy when it chooses, as in the definition of the true gentleman, "who is never rude except on purpose"; and the sad face of a juvenile, too curious to keep its head low as it sneaks along a grassy brake, is the first sign that the apes are moving out.

Gorilla gorilla goes away under cover of the bush, easing uphill and out across the old plantations and down again into dark forest of blue gum, leaving behind a spoor of fine, fresh droppings. Up hill and over dale comes *Homo* in pursuit, but *Gorilla* is feeling harassed now, and *Homo* is driven back from the forest edge by the sudden demonstration charge of a big-browed male who has been hiding in a bush. An oncoming male gorilla of several hundred pounds, with his huge face and shoulders and his lengthy reach, commands attention, and when the black mask roars and barks, showing black-rimmed teeth, we retreat speedily. The gorilla sinks away again into the green. To a branch just above comes a big sunbird with long central tail feathers—"Purple-breasted!" cries our dauntless birdman. *"That's* a new one!"

Finding their voice, the frightened guides yell at the gorilla, *"Wacho maneno yako!"* a Swahili expression often used to silence impertinent inferiors; loosely it means, "Don't give me any of your guff!" One of the Bashi, Seaundori, is scared and delighted simultaneously; grinning, he first asks eagerly if all had seen the charge of the gorilla, and then, imagining he has lost face by betraying excitement, he frowns as deeply as M. le Conservateur himself and fires nervously unnecessary orders. The visitors, too, are babbling in excitement; only the small Batwa trackers, grinning a little, remain silent. They follow the gorillas, never rushing them, just flicking steadily away with their old pangas in the obscuring tangle of lianas; even when the creatures are in view and no clearing is needed, the trackers tick lightly at the leaves as if to signal their own location to the gorillas and avoid startling them and provoking a panicked charge; this is the only danger from gorillas, which are as peaceable as humans allow. Though a leopard has been known to kill an adult male, the gorilla has no real enemies except for people. After years of protection at Kahuzi-Biega, two of the three gorilla troops that are more or less accessible have placed an uneasy trust in man's good intentions. For the second time they permit us to come within twenty feet before the bushes start to twitch and tremble, a sign that the ones still feeding in plain view might be covering for those that are withdrawing. And though our views are mostly brief, there comes a time when *Homo* and *Gorilla* are in full view of each other for minute after minute, not thirty yards apart. The apes are more relaxed than we are and also more discreet, since they do not

stare at our strange appearance; on the contrary, they avert their gaze from the disorderly spectacle that we present, lolling back into the meshwork of low branches and staring away into the forest as they strip branches of big leaves and push the wads of green into their mouths. At one point a dozen heads or torsos may be seen at once in a low tier of green foliage just below us; the black wooly hair is clean, unmatted.

Next morning we return to the gorillas' realm, taking a direction south and west and moving higher on the mountain. I had assumed that the small gorilla trackers were Batwa or Twa, the name used by Bantu speakers for all of Africa's small relict peoples, including the Bushman and the Pygmy. But these *pisteurs* are called Mbuti by the guides, and Adrien Deschryver (who was instrumental in creating the reserve at Kahuzi-Biega) later told me that they are apparently Bambuti or Mbuti hunters from the Ituri Forest to the north who were hunting gorillas in these mountains even before colonial times and maintain a small, separate village about five miles away from the nearest Bashi. Perhaps a certain mixing has occurred, for these little people are not "yellow," as the Mbuti Pygmies of the north are said to be, and may even be a little larger, though all three *pisteurs* are well under five feet. They have large-featured faces—big eyes, wide mouths, wide flat noses, big jaws—in heads that seem too big for their small bodies, but it is the way they act and walk that separates them most distinctly from the guides, for their bearing is so cheerful and self-assured that one is soon oblivious of their small size; they move through the undergrowth and with it, instead of fighting the jungle in the manner of the white people and the Bantu.

The trackers point their pangas at high forest to the south, consulting in a rapid murmur as they roll small cigarettes with makeshift papers. Then they set off up the mountain in a small-stepped amble that reminds me of the little Hadza hunters of Tanzania, checking gorilla droppings, following the gorilla paths in search of some fresh sign of feeding; a place is marked where the gorillas have exposed a large bed of small white woodland mushrooms, and these will be gathered on the return journey. Tambourine doves hurtle down the path, and from the forest all around come their long, sad-falling notes; we climb onward as a green-blue stretch of Lake Kivu comes in view, down to the east.

Mukesso stops short; he has heard limbs cracking. We hear nothing. But Mukesso is sure, and Kagwere and Matene do not doubt him; the Mbuti strike off into dense jungle, making no effort to keep down the noise, and have not gone a hundred yards when they cross the gorillas' path. The guides are nervous in this tangle, and even the trackers seem uneasy; they stop to listen every little while, ticking the vines and branch tips with their pangas. One whistles to the others, backs away a little; there is a big dark move-ment in the nearest bush, only feet away. We see the branches move, glimpse shifting blackness. Then the apes are gone, and the Mbuti do not follow; this place is dangerous, we must wait a little to see which way the apes will go.

Not so long ago, we had been told, a gorilla had killed one of the Mbuti and carried the body about with it for several days, but this exciting story is not true. It was a Bashi who was seriously bitten, not so long ago, when a panicked gorilla charged past him in making its escape, and this may account for the nervousness of the two guides. Most of the time Seaundori and the other guide, Rukira, are sullen and officious; no doubt they know that we don't feel we need them in the forest, for the *conservateur assistant* is not tactful with his staff, to put it mildly. To track gorillas, to hack paths through the forest is Pygmy work; since the guides carry no rifles like true askaris, they must know that they serve no purpose here what-ever. Like people all over Africa who have lost touch with the old ways, they live in mixed fear and contempt of the wild animals. Seaundori yesterday, Rukira today, were unnerved by the threat display of the great apes, and so today, to save face, they shout a lot of senseless orders and answer questions in querulous, aggressive manners.

Eventually, though we hear nothing, Kagwere jumps quickly to his feet and heads away into the forest, hacking and clipping, with Mukesso and Matene close behind. They trace an old path for perhaps a half mile, following it around

the east face of the mountain, pausing to listen, moving on again. The creatures are now well below us, working their way slowly up the hill; the Mbuti have anticipated their route of forage. We have only to ease along the mountainside, they will come to us. This bushcraft of the Mbuti is the way we had hoped to find gorillas. And soon the Bashi, growing bored, stop ordering us about, and even let us walk ahead so that we may observe the Mbutis' deadly tracking. . . .

Soon a young gorilla comes into view, climbing high into a tree. From a point a little farther on, a vast female is visible, sprawled in a comfortable crotch, in sun and shade, perhaps fifteen feet above the ground. Avoiding our stares she stuffs big, broad leaves into her mouth and pulls a thin branch through her teeth to eat the fresh light bark.

Slowly we sink down into the foliage. Through the winding light of the canopy the sky is blue, and to the nostrils comes the pungence of crushed leaves, the fresh green damp from this morning's rain, the humus smell of the high forest. Overhead a honeyguide and a tinkerbird sing fitfully; in the thrall of apes we pay them no attention. Observing the big female as she eats, a big male leans back into the vines on the ground behind her; probably he is too heavy now to climb. And seeing his vast aura of well-being, one understands the African theory that the gorilla was formerly a villager who retired into the forest in fear of work.

Young gorillas come, still curious about the forest; they play with each other and with the trees, using their opposed toes to brace their climbing. One juvenile is lying belly down over a branch, all four limbs dangling; he rolls over and down, to hang by one hand in the classic pose and scratch his armpit. He has wrinkled gray bare fingers and gray fingernails. Briefly, he roughhouses with an infant, who flashes a small pale triangle of bare rump, and the Mbuti laugh, mopping the sweat from their wet faces with handfuls of fresh leaves. For a time the young ape hangs suspended by both arms like a toy gorilla; lacking the discretion of his elders, he leers at man in a thin-lipped, brown-toothed grimace that matches his brown eyes, those eyes with the small pupils in the flat and shining gaze that does not really seem to see us. The gorilla face looks cross and wild and very sad by turns, though scientists tell us that all primates but ourselves are incapable of emotive expression.

To sweet-scented dung, like rotted flowers, comes a yellow butterfly; somewhere unseen the flies are buzzing, and a tambourine dove calls. From the undergrowth come deep contented grunts, then vast stomach rumblings and the sharp crack of a branch that does not break the rhythmic sound of the female's chewing.

Soon the last of the gorillas has swung, climbed, lolled, and cleaned its bottom, beaten its chest in a series of soft tappeting thumps, lowered to the ground the bellyful of vegetation that makes gorilla legs look small and thin, and vanished once more into the green. We have watched them for an hour, and we are delighted; we talk little, for there is little to be said. On the way home the Mbuti cut themselves packets of bark strips for making bush rope and gather up the small white mushrooms of the forest.

PART 3

1979, 1986

In a clear August dawn of 1979, as the sun was born again from the Indian Ocean, a silence arose like memory from the turning Earth and with it a promise and elation. The shining wing crossed the equator at Mount Kenya, passing over into Tanzania as Kilimanjaro and Mount Meru rose like black islands from a sea of clouds. Tomorrow or the next day I would leave on a month's safari into the Selous Game Reserve in southeastern Tanzania, said to be the greatest stronghold of large wild animals left on Earth. Its area of 22,000 square miles makes it the largest wildlife sanctuary on the continent, almost three times the size of the great Serengeti National Park, yet of all the great parks and game reserves in East Africa, it remains the least accessible and the least known. The safari, led by Brian Nicholson, who had served as a game warden of the Selous for more than twenty years, would proceed to a remote point at the confluence of two great rivers in the far south of the Selous, where Brian Nicholson and I would cross the rivers and penetrate as far as possible on foot.

Besides the oil crisis, Tanzania was suffering from a year of failed crops and the attrition of its eight-month war to depose Field Marshal Idi Amin Dada of Uganda. The many Tanzanian soldiers still in Uganda had completed the butchery, begun by the fleeing Ugandans, of the wild animals in the great parks, selling meat and skins as well as ivory. Meanwhile, Kenya, was struggling to recover from the epidemic poaching of the 1970s, which had threatened the existence of such splendid species as the black rhinoceros, reticulated giraffe, and Grevy zebra. In the light of all this sickening news, and of the chronic political instability that clouds the future of Africa, "the last safari into the last wilderness," as its sponsor, a young British member of Parliament, described it, did not seem such a fanciful description after all.

Northern Selous

On 21 August we flew inland over the cloud shadows and small settlements of the coastal bush, entering the northeast corner of the Selous. The plain beneath had been turned hard black by recent burning. The warden, seated in front next to the pilot, shook his head. "Poachers, I should think. Always had a problem with them in the north because here the villages come up so close to the boundaries. I've seen snares strung through here for almost fifteen miles, and dead animals rotting all along the way." Off to the west rose the hills of the central African plateau, and soon the plane was crossing the Rufiji River. It was now midafternoon, and large groups of elephant and buffalo were moving peacefully toward the shining water.

The night before, as if to signal our return to the African bush, hyena and lion howled and roared, though not with laughter, and toward midnight hippopotami resounded from their pools deep in the Kingupira Forest. The wistful birdcalls of the African night died one by one; soon the ring-necked and red-eyed doves began to call, and then the tiny emerald-spotted wood dove with its sad, descending coos, and the dawn scream of a fish eagle, the tinny notes of the trumpeter hornbill, the nasal, jeering squawk of hadada ibis—as the canvas filled with light, I lay on my camp cot in the crisp green tent in the greatest happiness.

Where a new flush of green had arisen from the burned-out black was a herd of Nyasa wildebeest, larger, paler, and more handsome than the race on the north side of the Rufiji, with roan flanks and haunches where that animal is gray, and a remarkable white blaze across the forehead. Farther west, in dry *miombo* woods, a fine big civet cat started from a clump of tawny grass, then moved away a little distance before stopping to turn and have a look at us. The civet was black-faced, lustrous in the sere pale grass, averting its head just a little, the better to listen, and going on again when it heard no more than the soft vibration of the Land Rover. The civet is not a cat at all but a large omnivorous relative of the mongoose and the honey badger. It eats fruit and carrion as well as small animals and birds, and helps to propagate the fruit trees that it frequents by depositing their seeds in the defecation place that it returns to again and again, sometimes for years. There the seeds thrive, not only because of the powerful fertilizer but because the small rodents that normally eat up the fallen seeds avoid the civet smell and leave the tree nurseries alone.

      A group of buffalo went rocking away through the small trees, a lone hyena sat up like a sphinx. On the savanna as well as in the open woods there were impala, which seem to occupy the ecological niches filled farther north by the gazelles. In the Selous, the impala is the major prey of the wild dog and the scarce cheetah. In the long grass of the *miombo*, these elegant antelope have the kongoni habit of climbing on to ruined termite mounds to scan the landscape. . . .

With the photographer Hugo van Lawick, I spent a day at a large pan called Namakamari, or the Catfish Pool, a harmonious place set about with small black cassia trees in yellow blossom. . . . A flock of thirty-two openbill storks, with heavy flapping, settle in a sepulchral arrangement on the bare limbs of a dead tree; the openbills are named for the odd space between their mandibles, through which one may see the blue African sky. As the sun rose, the hippos, which had settled somewhat at our approach, lowered themselves deeper still into the thick gray-brown broth of their own making. A gray heron in their company was evidence the water could scarcely be deep enough to immerse a standing hippo, far less a swimming one; the enormous animals were resting on their knees.

      I sat very still in the thin shade of a tree that grew from an ancient termite hill close to the shore. On one dead limb over the pool, two hammerkops peeped sadly as they mated; a pygmy kingfisher, turquoise and fire, zipped into a burrow hidden in the mound behind me. A *brrt* of wings preceded the arrival of chinspot flycatchers, and soon other birds came to the bare limbs and dead snags nearby: doves and rollers, a white-headed black chat, the lesser blue-eared starling, sparrow weavers, and a brown-headed parrot that could not make up its mind whether it should investigate or flee. Striped skinks emerged beside my book, and the parrot followed me all around the little hill, clambering along on the limbs over my head with electric shrieks of indignation as I stalked a small deliberate slow bird, olive-gray above with pretty gray bars on a white breast, called the barred warbler. Searching for mites, the warbler worked from the base of a small bush up to the top, flew down and started again, always moving upward from the bottom, until it had circled the mound to my place once more, where it gleaned the leaves near my right hand.

      A herd of impala picked its way around the pool to a point just yards from where I sat. Their harsh tearing snorts as they suddenly departed would warn me, I thought, of the approach downwind of any lion. Soon warthogs came in from the far side, progressing forward on their knees, tails whisking and manes shivering as they snouted and rooted in the earth. From the pond, in the thick heat of the growing morning, came a pungent duck-pen smell to which the Egyptian geese that swam around at the edges of the hippo herd made only a pitiable contribution. Seemingly content with the sheer overwhelming presence of their huge and indelicate companions, the geese never appeared to feed, but simply stayed close; they retreated

only when washed backward, attending minutely to each thrash and heave as the herd barged about in its small space, as if a goose had much to learn from hippopotami. The cacophony of groans and blares, snorts, puffs, and sighs would subside with the submergence of the mighty heads, leaving only a mute cluster of shining wet boulders on the still surface of the pool. Then, one by one, the heads protruded, froggish pink eyes and round pink ears, followed by the generous nostrils that can be closed tight under the water.

Sometimes hippos remain beneath for minutes at a time, thinking long thoughts or cooling the cumbersome machinery of their brains, or—in deeper water—enjoying a short stroll over the bottom. But in these close quarters the commotion resumed rapidly, a quake and rumbling from beneath the surface, then a roar and wash as the huge bodies surged, and way was made for two pink-eyed gladiators, which draw near slowly, splitting each other's ears with heavy bluster. Sometimes one will turn aside, not to flee but to hoist its hind end out of the water long enough to defecate, the fleshy furious short tail whisking muddy manure into the unoffended face of its assailant. Since subsurface elimination would be more relaxing, Hugo has concluded that this strenuous act—surely an aggressive one among human beings—is a gesture of submission among hippos.

Many of the outbursts were not true fights but the threat display of a female hippo, directed at those who approached her calf too closely; this maternal solicitude never failed to incite an uproar, which soon deflated into peevish snorts, as if to say, "What can be done with such crude people!" Since the animals were all piled up together, the cow appeared to be drawing a fine line, but no doubt she could perceive a threat not discernible to the casual observer. Despite appearances, hippos are sensitive and easily upset; they were not reconciled, even hours later, to the presence of Hugo's car, which they stared at all day with suspicion and pursued with bluster charges whenever it appeared to be departing, in order to speed it on its way.

It was noticeable, however, that when real fights occurred between two males, the herd did not join in the uproar but fell silent, as if watching carefully for a sign that the hippo hierarchy was about to change. Even the Egyptian geese retired as the gigantic creatures reared up on their hind legs, mouths wide and ivory clacking. Their huge heads locked, the titans twisted, crashing back into the water in an attempt to come to grips as a dung-filled wave rolled across the pool, flushing the birds up from the margin and washing the water lettuce with a rich soup of manure. Then a third male came in from the side, in discreet silence, to deliver to one of the straining contestants a terrific bite upon the flank, driving it off. He then turned upon the other and engaged it in a contest of jaws, which he soon won. Only when the fight was over did the nervous herd release its tensions with a mighty uproar, as if the opinions of each one had been vindicated. Most of this was ritualized combat, minimizing injuries, as it is among many if not most of the horned and antlered animals, but hippo bulls may be slashed open by the enormous shearing teeth, and often die. At midday one of the vanquished, apparently banished from the pool, came very quietly out of the hot scrub, anxious to get in out of the heat; he stood indecisively on the bank, great head resting humbly on the mud, as if listening for favorable vibrations. If so, he heard none and decided not to risk it, for after a while he turned away and walked back slowly into the bush, revealing a large open gash on his hindquarter.

Wild dogs visited the pool, first two, then the whole pack. The strange bat-eared creatures circled around behind the car with curiosity, emitting that odd grunt-bark of alarm that contrasts so strangely with their birdlike twitterings of greetings and contentment. These were all good-looking animals, with shining black masks and brindle on the nape and shoulders, glossy black and yellow-silver bodies, irregularly splotched, and alert clean white-tipped tails. All the large carnivores we had seen so far in the Selous—the hyena, lion, and wild dog—were big healthy animals with fine coats, entirely lacking the scuffed and tattered character they display elsewhere. This may be because the abundance of water and good pasture reduces the need of the prey animals for seasonal "migrations," and the resultant stress of leaving their own territories.

In the late afternoon the hippo calves began to surface, the small heads appearing right beside their mothers. The calves are born and suckled in the water, and can lie so low, with only their nostrils emerging, that we did not realize how many there were here. In the scrub on the south side of the pool I surprised a hare; it ran off from its dreaming place beneath my feet, the low sun shining through its ears. Nearby, a leaf-nosed bat hung in a low thorn bush like a dried fruit.

Mbarangandu River

It was close to noon on 26 August when we broke camp and headed south. . . . From the southern distance, under odd round-topped hills, rose the white of broad sand rivers, then the glint of water as the gray and ghostly stubble of *miombo* scrub, high on the ridges, turned into red, green, and copper woods. Where the track descended the last slope of the rivers, the woods on both sides were littered with fresh buffalo dung.

The Mbarangandu rounds its final bend under steep bluffs on this north side, where the ridges level out onto the plain; on the south side, across the river, lie grass banks and brakes, then open woodland that soon begins to climb into the hills that separate the Mbarangandu from the Luwegu.

On broad deltas of white sand numerous water birds flew back and forth about their business, and bands of waterbuck lay on the margins like the tame and stately park deer of old paintings. Beyond the tall borassus palms on the far side of shining waters and white sands rise the blue hills: from this place, the Selous extends more than one hundred miles to the southwest beyond those Mbarika Mountains, but there are no tracks beyond this point of rivers.

The warden stretched out in his camp chair with the greatest satisfaction, a rare grin breaking out upon his face. "You're a long way from anywhere now, I can tell you! The Selous is the finest wildlife habitat in Africa, and the Mbarangandu is the heart of it!"

Before sunset, the diurnal birds were still; only the fish-tailed drongo was still flying. The water dikkop sang its sad, descending song of twilight, and nightjars left their camouflage of bark and leaves to settle on the warm sand of the tracks. At dusk, the tiny scops owl began its trilling, and toward midnight, a fish owl at the water's edge not far upriver gave a strange, low, and lugubrious grunt that was heard occasionally throughout the night. After two days of human presence, this shy bird was not heard again.

Because it was a lovely afternoon, we went on for a mile or more, wading upriver. Far upstream at the next bend, an elephant moved peacefully, walking on water. "Not bad ivory, that!" Brian remarked, then adjusted his enthusiasm quickly. "Nothing special, of course, not much more than fifty, I should say."

Waterbuck nuzzled the green haze appearing on the damp flats of the river, and two lionesses rose from the far end of the flats and trotted to the high grass of the riverbank as Simon cried out, *"Simba! Simba!"* At the mouth of a small river of white sand that came in from the north, a heavy lion spoor included many neat prints made by cubs; there were also the fresh pug marks of a leopard. Not far up the sand river, feeding placidly in rank green swale, was an elephant cow with a juvenile and a young calf, and Brian, seeing calm elephants close at hand for the first time since he had arrived in the Selous, sat down on a high bank and watched them in contented silence. He is obviously very fond of elephants, and, observing them, he permitted himself a fleeting smile, pursuing some elephant reverie or other.

After some minutes, the cow's trunk stiffened as she got our wind; the trunk rose slowly. Then she hurried her young into the cane at the edge of the riverain thickets, and in moments the vast animals were gone.

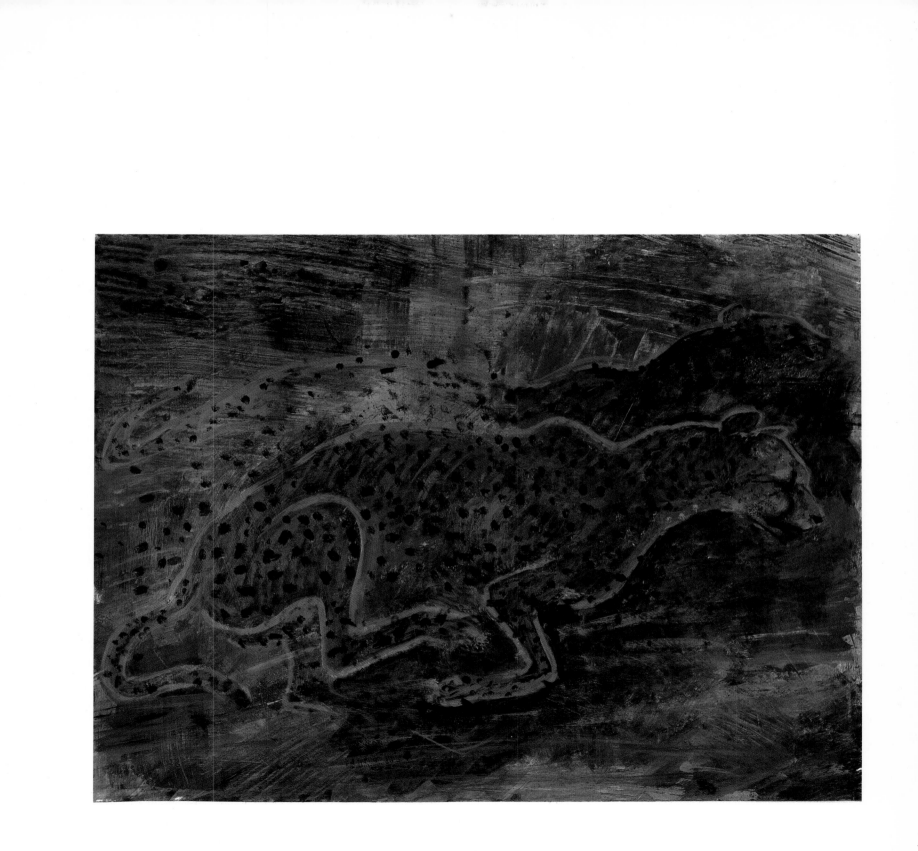

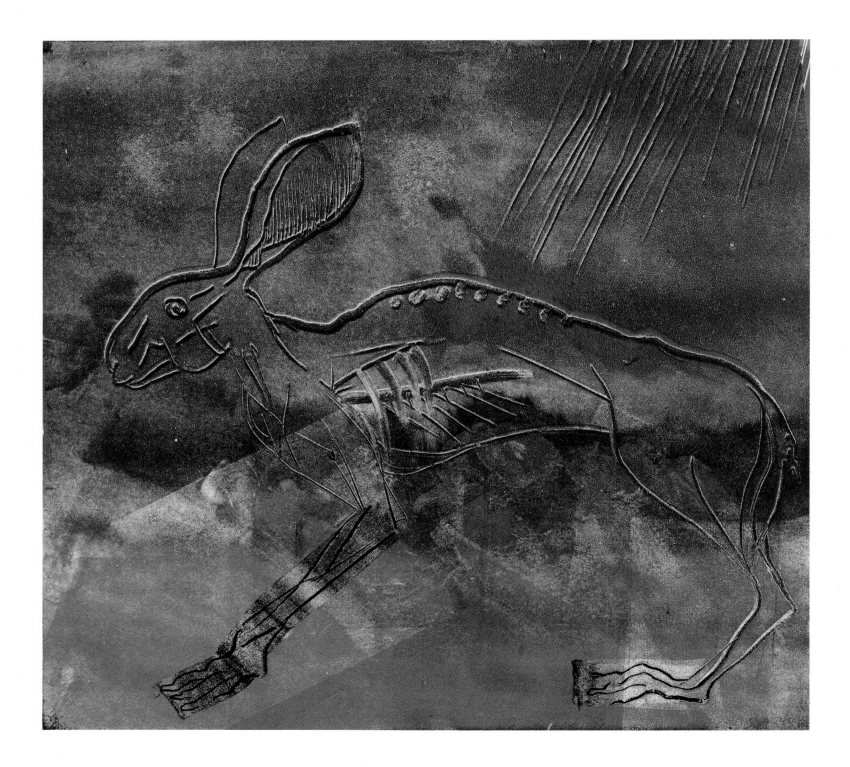

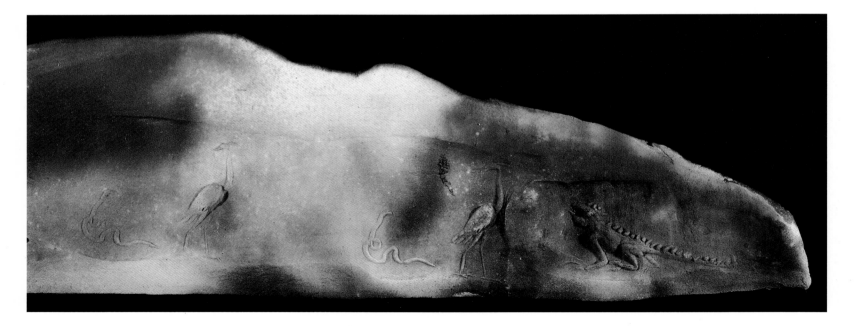

Barefoot in cool shallows on clear sand, we wandered upriver. A flock of green sandpipers, twelve or more, was feeding along the edges of the bars; this was a migratory flight south from Eurasia. Sneaking up close to see them better, I was startled by the explosion of a hippo from a small side channel near the bank, not twenty yards away. At this range, one is very much aware that excepting the elephant and the white rhino, the hippo is the largest land animal on earth, and since I was between it and the main channel, I was glad that this one was aware there was no deep water left in the Mbarangandu; it made for the thickets, which closed discreetly over its huge shining hindquarters.

In contentment, we strolled slowly back along the south bank of the river, pausing to investigate the twin scrapes that the rhino makes by rooting and kicking while scattering its dung. Africans say that God sewed skins on all His animals with a big needle and, becoming tired, asked the rhino, which was last, to finish up; the rhino did not make a good job of it, which is why it has so many loose folds and wrinkles, and furthermore, it swallowed the needle with the job half-finished, which is why, still searching for the needle, it insists on scattering its dung.

The hippo does the same, though its reasons differ. When it asked God if it might take up residence in the water, God refused, declaring that an animal so enormous would eat up all His fish and water plants and spoil His rivers. The hippo promised it would feed on land, and to this day waggles its dung about on bushes just to prove to God that it is going properly about its business.

Just after midday, along the wood edge, we encountered a fine sable antelope grazing not fifty yards away. Because the wind was in our favor, the big black animal did not scent us, though he reared his harlequin head as if to listen. Soon he was joined by three more bulls that stepped one by one across the track; one was still young, chestnut-colored and slighter in horn and body than the others. The animals grazed peacefully on the bright green tufts under the fire-coppered leaves of a blackened rain tree. Then the air shifted, the armed heads jerked to taut attention, the white-blazed faces turned to stare even as the shining bodies gathered and sprang away across the hillside, whirling up the dust from the black ground. And as they streamed between the trees, it was easy to see why the sable (and the roan) were named *Hippotragus*, "horse antelope," not only because of their size and strength but because, unlike such peculiar relatives as the wildebeest and the kongoni, they move with the elegance of horses, heads high, chins toward their chests, as if to accentuate the grand sweep of the curved horns. There is none of that odd bouncing gait, called "pronking," that is seen in lesser antelopes, even the gazelles. That morning an impala had pronked away into the woods in inappropriate response to the sudden appearance of our blue Land Rover, and kongoni tend to pronk as a matter of course. This foolish gait seems suited to the long-faced kongoni, with its striped ear linings, blond hindquarters, and rumpled horns; at times, losing all forward motion, it appears to bounce straight up and down.

The wildebeest is also a born pronker. On the ridge flats, in low scrub, more than one hundred of these flighty animals came together, jostling and crowding one another as they pushed forward for a better look at man. The lead bull had imposing horns, which glinted in the sun like horns of buffalo, but such horns are ill-suited to a long sad face with peculiar ginger eyebrows. The wildebeest has a goat's beard and a lion's mane and a slanty back like a hyena; the head is too big and the tail too long for this rickety thing, and Africans say that the wildebeest is a collection of the parts that were left over after God had finished up all other creatures. . . .

One day down by the Mbarangandu, a lone bull elephant was seen wandering slowly back and forth along the river, as if it had lost its last sense of direction; when Hugo approached, it actually drew near the car—"but not at all in an aggressive way," he said. "It had what looked like a spear wound on the side of its face, and was holding its ears back tight against its neck in a

strange manner, and it seemed to me that its jaw was swollen. Then it wandered away again to a small pool where it sprayed a little water on its head and beneath its ears."

That night or early the next day, the elephant sagged down and died against the green grass bank between the plain and the white sand of the river, and a day later, more than three hundred vultures had assembled, including two beautiful palm-nut vultures, which may have joined the madding throng for social purposes, since they are not known to consume carrion. The first at the feast shared the carcass with hyena and lion, but perhaps these animals were already well-fed, for as the hordes of dark birds circled down out of the sky the carnivores withdrew, and the elephant disappeared beneath a flopping mass of vultures that stained the river sands all around a dark gray-green.

The elephant carcass was inspected before the birds reduced it to a cave of bones, and as it turned out, the left ear it had held so close was protecting a great infected wound that maggots had eaten out down to the bone; there was also the separate wound on the left side of the head, apparently inflicted by a spear. The elephant's putrefying wounds were some weeks old. The game scouts say that local poachers of the region don't use spears, only arrows tipped with *Akokanthera* poison and old musket-loaders armed with poisoned shot; they thought the larger hole looked like a musket wound.

But the tracker Kirubai thought that the two wounds were caused by tusks of another elephant; sometimes, he said, bull elephants will fight so violently that tusks are broken, and an elephant may wander around for months with such a wound before it dies. As a former poacher, he had not seen much evidence of poaching, at least not here in the far south of the Selous; he doubted that poachers would have overlooked the five valuable tusks we have found along the way.

Although buffalo and elephant were here at Mkangira when we arrived and a rhino was seen on the second day, these large beasts soon vanished from the region. Unlike park animals, they avoid the presence of man. But after the middle of September, as if anticipating our departure, a herd of several hundred buffalo came down to the south bank of the Mbarangandu, and the next day two rhinos appeared in the same place; elephants reappeared on the northern plain, and a loose herd of eight bulls, including one with a large single tusk, could be seen each day. Discovering the dead one by the river, these elephants stayed near it for a day or two in answer to some elephantine instinct, perhaps more akin to respect for death than man may think, although the dead kinsman was now no more than a hollow gray mound of hard-baked skin, a sagging armature of bone.

Almost every night the restless lions could be heard on both sides of the river, and sometimes leopards, and invariably hyenas. Because of the smell of the buffalo and impala that were killed every few days to feed the camp, the hyenas were bold nightly visitors, skulking about the kitchen area and between the tents, leaving behind the strange long prints that like the rest of their appearance is more suggestive of the dog than of the aberrant cat that they really are. One night another hyena clan made its own kill on the far side of Luwegu, filling the night with excited whoopings that turned to eerie giggling and laughter. Out there in the dark where the hyenas were tearing the wide-eyed victim into pieces, those crazy noises would be ringing in its ears.

The brown flood sparkling under the moon was perhaps two hundred yards across, yet it was shallow enough for a man to wade the chest-high water were it not for the big crocodile that had shown itself now and again in recent days and taken most of the languor out of bathing. Between birdcalls, in every silence, came the soft wash of the two rivers, pouring away to the north and west to meet the great Kilombero that comes down out of the Nyasa Highlands on its way to the Rufiji and the sea.

Southern Selous

On an early African morning, Brian and I set off for the south, wading across the Mbarangandu not far above its confluence with the Luwegu. Climbing the ridges between rivers, we shall follow the game trails for about eight miles, then descend to the Luwegu and continue south for perhaps three days before turning east to explore a high plateau with its own extensive swamp or pan. From the plateau, we shall descend a tributary river that comes down off the west escarpment of the plateau and turns northeast, arriving eventually at the Mbarangandu. "You and I will be the first into most of the country that we're going to cross," the warden told me.

By early afternoon the clouds have gone, and the day is dry and hot. Drinking gratefully from the brown river, I realize how rare now are the places left in Africa where one can drink the water without risking bilharzia or worse; in the Selous, one can sip with impunity from pools and puddles and even from big footprints in the mud. Later I find a safe bathing place behind a silvered log, and lie back for a long time in the warm flood, watching the western sky turn red behind a gigantic baobab across the river. Behind me in the forest, an elephant's stomach rumbles—or perhaps the elephant is pondering my scent, for Brian says that what is usually called stomach-rumbling in elephants is actually a low growl of apprehension and perhaps warning. Hornbills gather in the mahoganies over my head, and I am attended by a small dragonfly, fire red in hue, that might have flown out of the sunset. . . .

A half moon rests in the borassus fronds, and a tiny bat detaches itself, flits to and fro, and returns into the black frond silhouette. We lie peacefully upon our cots and watch the stars. From the forest comes the hideous squalling of frightened baboons trying to bluff a leopard, or so we suppose, since leopard sign is plentiful around the camp.

In the moonlight the bull hippos of the herd move in close to the bank to bellow at our fire; in trying to frighten us away, they panic one another and porpoise heavily away over the shallows, causing great waves that carry all the way across the river and slap onto the mud of the far bank. Man does not belong here, and the hippopotami cannot seem to accept us; we have disrupted their whole sense of how the nighttime world unfolds in the Old Africa. They do not go ashore to feed but remain out there just beyond the firelight, keeping watch on the intruders and banishing our sleep with outraged bellows.

Leaving the river forest in the early morning, we emerge onto an open plain and head south again over low hills between broad bends of the Luwegu. Shadows deep in the scrub ahead have shifted, and soon a bull elephant moves out into the open, in no hurry, since the tracker Goa has left him time and space in which to take his leave. There are more elephants during the morning, in twos and threes and fives; although on foot, we are rarely out of sight of them.

Behind the elephants is a large grove of borassus palms, with their graceful swellings high up on the pale boles. At this season, the borassus carries strings of fruit like orange coconuts, which are sought out by the elephants. Here and there in the dry hills, far from the nearest palms, lie piles of dried gray borassus kernels, digested and deposited, from which the last loose dung has blown away.

A solitary eland bull, a glimpse of kudu. The country is more open here than it is to the north of Mkangira, and the racquet-tailed roller of the *miombo* woodland has been replaced by the lilac-breasted species of the savanna. Even the tsetse seem to have lost the appetite they show in the closed woods; I watch them alight on the shirt of the man ahead, but they do not bite.

Against the blue hills to the west stands a cow-calf herd of elephant. Getting our scent, a young cow leads the juveniles away while the old matriarch stands guard, trunk high, as if in warning, and soon we see the young toto hastening away after the others, the top of his small earnest head scarcely visible in the high grass. For a long time a gray eland bull stands watching us, attended by three soft brown cows with calves. Like other striped animals—kudu, bushbuck, zebra—eland are taboo to the tribesmen to the east of the reserve, who know that eating striped beasts may bring on leprosy. (This view is shared by tribesmen of the Sudan-Zaire border, who will not eat bongo.)

In numbers of animals seen this morning, the far south of the Selous compares with the great parks, but the tameness of almost every creature in this country south of Mkangira has nothing to do with the aplomb of sophisticated park animals, which, being resigned to the human presence, are not tame so much as half-domesticated. Here the confiding curiosity appears to stem from a trusting innocence of man unlike anything I have seen elsewhere in Africa. . . .

Goa follows the old paths of the elephants, which follow the ridge lines and avoid the stony depths and thorn of the karongas; it is pleasant to suppose that most of these clear trails, two feet across, from which all grass has been worn away, have never been seen or walked by human beings. Eventually a path descends toward the river, and from a bluff we contemplate an elephant with calf, at rest on a clear shoal in the middle of the glittering Mbarangandu; in the new light of the river morning, the creatures seem to dream, lulled to forgetfulness of where they might be going by the clear torrent from the

southern mountains that casts sparkling reflections up the gray columns of their legs. When the cow and calf move on, we wade out into the sun-filled flow and continue southward.

A big pan not far away up the east bank of the river has always attracted elephant and game; in other days, this pan held a large pool that was permanent home to crocodiles and hippos, but the hippos wore a runway to the river through which at last the whole marsh emptied out, obliging both aquatic species to gain a living elsewhere. It was here at Likale, Brian says—and he points to an open wooded hill on the south side—that he saw the greatest and most splendid kudu of his life. On the north side, a number of elephants are in sight, including two that cross the river to join those on the western bank. In the dry pan, one hundred buffalo stand in a compact herd, with nearly that number of impala, thirty warthogs, a band of kongoni with a new calf, a zebra herd at the far wood edge where had been seen the great kudu of yore, and a quorum of the elongated birds that stand about at the water's edge all the world over.

We wade the river and make camp just opposite the pan. To the east, the land rises to steep red escarpments; these broad valleys with steep cliffs may well be indications of ancient fault lines of the Great Rift that runs south from the Red Sea to the blue Zambezi. On all sides is an airy view of the Old Africa, and I am delighted that Brian wished to camp here and sorry to see the look of discouragement upon his face. "I'd been looking forward to this place," he says, "and it's the best we've seen, yet it just doesn't compare with the way it was. The buffalo herd used to be three or four times that size! And we have yet to see a single elephant with decent ivory!"

I repeat my arguments, to reassure myself as well as him: the scarcity of big bulls with large tusks is bothersome, of course, but it is simply not possible that the thousands of elephants observed from the air in 1976 and 1978 could have been killed off by poachers or even by plague, since on this trek south of Mkangira we have found not one dead elephant. Also, the fact that the groups were so small was strong evidence that they were not being harassed. And wasn't there also a certain negative reassurance in the case of the rhino? Regularly along our way we have come upon fresh rhino scrapes, and since the rhinoceros regularly returns to the same place to defecate, almost all of these scrapes must represent different animals: yet the pair this morning were the first we have actually seen on this safari.

In the 1960s a number of rhino were killed because of a notion among Orientals that the compacted erect hair of the rhino "horn" was a cure for impotence and certain fevers: the accelerated rhino slaughter of today has come about because rich Arabs of the Middle East have made a fashion of daggers with rhino horn handles, for which they are willing to pay more than six thousand dollars apiece. The fad or fetish for these phallic daggers has jumped the already very high price of horn up to five thousand dollars per kilo in Hong Kong—the worthless stuff commands more than pure gold—and unless drastic measures are enforced, and soon, an ancient species may vanish from the earth millions of years before its time because of sexual insecurity in *Homo sapiens*.

Even ten years ago one could take for granted encounters with a few rhinoceros in East Africa, these days it is a stroke of luck to see one. So far as is known, black rhinos have been all but eliminated from Uganda. Kenya's recent population of fifteen to twenty thousand rhino has been reduced to between twelve and fifteen hundred. Since the early 1970s, the rhino in the Tsavo parks have declined from seven or eight thousand to one hundred eighty; in the small Amboseli park, the decline is from fifty to ten. Rhino poaching has crossed the Kenya border from the Maasai Mara into the Serengeti, and the other important Tanzania parks—Ngorongoro, Manyara, Tarangire, Ruaha—have already lost at least three-quarters of their populations. In the Selous the most recent estimate of rhino numbers, made during the air survey of 1976, arrived at the figure of four to five thousand, which must be the last large healthy population of this species in the world. . . .

Dark clouds and wind. At dusk, under the eastern bluffs where an elephant is throwing trunkfuls of fine dust into the air, a ghostly puff of light explodes, another, then another. I cannot see the elephant, only the dust that rises out of the shadow into the sunlight withdrawing up the hill. At dark a hyena whoops and another answers, for the clan is gathering, but their ululations are soon lost in a vast staccato racket, an unearthly din that sweeps in rhythmic waves up and down the river bars, rising and falling like the breath of earth—then silence, a shocked ringing silence, as if the night hunters have all turned to hear this noise. Somewhere out there on the strand, I think, a frog has been taken by a heron; my mind's eye sees the long bill glint in the dim starlight, the pallor of the sticky kicking legs, the gulp and shudder of the feathered throat. The frog's squeak pierces the racket of its neighbors, who go mute. But soon an unwary one, perhaps newly emerged from its niche under the bank, tries out its overwhelming need to sing out in ratcheting chirp; another answers, then millions hurl their voices at the stars. The world resounds until the frogs' own ears are ringing, until all identity is lost in a bug-eyed cosmic ecstasy of frog song. In an hour or two, as the night deepens, the singing impulse dies, leaving the singers limp, perhaps dimly bewildered; remembering danger, they push slowly at the earth with long damp toes and fingers, edging backward into their clefts and crannies, pale chins pulsing.

Toward midnight I am awakened by a bellow, a single long agonized groan; a buffalo has cried out, then fallen silent. Perhaps something is killing it, perhaps a lion's jaws have closed over its muzzle, but I hear no lion, now or later. At daybreak a birdcall strange to me rings out three times and then is gone, a bird I shall never identify, not on this safari nor in this life. As a tropic sun rolls up onto the red cliffs across the river, setting fire to a high, solitary tree, the moon still shines through the winged piliostigma leaves behind the tent. . . .

On a grassy hillside of small trees, a burst of terminalia saplings has sprouted out of a rhino scatter on the path, and there is lion spoor. Then Goa's hand is up: he points. A large dark animal stands in the high grass below the ridge line. With binoculars I pick out four more sable, lying in the copper-colored grass of the ridge summit; they do not see us but face eastward, over a broad sweep of the river, two miles away. Then another big male, long horns taut, is standing up and staring in our direction, his shiny black hide set off by the white belly, chestnut brow abristle with morning light above the twin blazes of his face. Now another bull jumps up, and then another. The bulls regard us for a little while before leading the herd away along the ridge line and down on the far side. A dozen animals cross the sky, including a young calf, and probably as many more never emerged from the high grass but simply withdrew down the north side of the ridge. The two lead bulls watch the others go before moving up through a small glade of silver trees at a slow canter, turning against the sky to stare again, then vanishing.

On an elephant slide on the north end of the plateau lies an old cracked tusk that has been there for long years— mute evidence that no man comes this way. Descending, we head north through rough and unburned country to Mto Bila Jina, the River Without Name. Toward sunset, where the river rounds a bend, three buffalo bulls stand together at the end of a long stretch of clean white sand. Brian is footsore and discouraged and irritable, and he knows that for the first time on our safari, the morale of the Africans is precarious due to the real or imagined need of meat. Earlier he had said that in shooting a buffalo too much of the meat would be wasted, even if two porters gave their loads to the others in order to lug all the meat that they could carry, but now he decides without further ado to execute one of the bulls. Leaving the rest of us behind, he stalks with Goa to a point on the stream bank not twenty yards from the three buffalo below, a point-blank range from which he is sure to drop the animal with the first bullet. At the crack of the rifle, the buffalo sags down upon the sand with the windy groan of death, and in the echo of the shot, the Africans laugh and clap their hands together. He is restless, unable to make his peace with it; I had not suspected that he would be so upset. After all, this man has killed thousands of animals, and no doubt hundreds of buffalo among them; the buffalo is not an endangered species, and these three bulls may have passed the reproductive age since they had wandered off from the large herds. Even the "waste" will not be seen as such by the carnivores and vultures that will reduce this buffalo to bones in a few days. It is not the waste that bothers him, but the intrusion by man into the "heart of the Selous," which was symbolized by that isolated shot. . . .

A cold clear morning. Well before daybreak, voices murmur and human figures move about, building up the fire to keep warm. A smell of carrion hangs heavy on the air, but the leopard, heard again last night, has not visited the buffalo, nor did the lions follow up the circling vultures.

In the cold sunrise, as we depart, a stream of parrots in careening flight recaptures the sausage tree across the river; the fleeting human presence will lose significance with the last figure that passes out of sight in the dawn trees.

We are headed north again, into burned country, and the spurts of green grass in the black dust are sign that these fires preceded those we made on the way south. In the bright grass, the animals are everywhere, making outlandish sounds as we approach; the kongoni emit their nasal puffing snort, the zebra yap and whine like dogs, the impala make that peculiar sneezing bark. But two buffalo that canter across our path are silent, the early red sun in the palm fronds glistening on their upraised nostrils, on the thick boss of the horns, the guard hairs down their spines, the flat bovine planes of their hindquarters.

At the edge of the plain, between thickets and karongas, Goa rounds a high bush and stops short Without turning around he hands the rifle back, as Brian and I stop short right behind him.

In a growth of thin saplings, at extreme close quarters, stands a rhinoceros with a small calf at her side. The immense and ancient animal remains motionless and silent, even when the unwarned porters, coming up behind, gasp audibly and scatter backward to the nearest trees. Goa, Brian, and I back off carefully and quietly, without quick motion: I am dead certain that the rhino is going to charge, it is only a matter of reaction time and selection of one dimly seen shadow, for we are much too deep into her space, too close to the small calf, to get away with it. But almost immediately a feeling comes—a knowing, rather—that the moment of danger, if it ever existed, is already past, and I stop where I am, in astonished awe of this protean life form, 600,000 centuries on Earth.

In the morning sun, reflecting the soft light of shining leaves, the huge gray creature is magnificent, the ugliest and most beautiful life imaginable, and her sheep-sized calf, which backs up into her flank, staring intently in the wrong direction, is of a truly marvelous young foolishness. Brian's voice comes softly, "Better back up, before she makes up her mind to rush at us," but I sense that he, too, knows that the danger has evaporated, and I linger a little longer where I am. There is no sound. Though her ears are high, the rhinoceros makes no move at all, there is no twitch of her loose hide, no swell or raising of the ribs, which are outlined in darker gray on the barrel flanks, as if holding her breath might render her invisible. The tiny eyes are hidden in the bags of skin, and though her head is high, extended toward us, the great hump of the shoulders rises higher still, higher even than the tips of those coarse dusty horns that are worth more than their weight in gold in the Levant. Just once, the big ears give a twitch; otherwise she remains motionless, as the two oxpeckers attending her squall uneasily, and a zebra yaps nervously back in the trees.

Then heavy blows of canvas wings dissolve the spell: an unseen griffon in the palm above flees the clacking fronds and, flying straight into the sun, goes up in fire. I rejoin the others. As we watch, the great beast settles backward inelegantly on her hindquarters, then lies down in the filtered shade to resume her rest, her young beside her.

Yet seeing the innocent beast lie down again, it was clear how simple it would be to shoot this near-blind creature that keeps so close to its home thickets, that has no enemies except this upright, evil-smelling form, so recent in the rhino's world, against which it has evolved no defense. Its rough prong of compacted hair would be hacked off with a panga and shoved into a gunnysack as the triumphant voice of man moved onward, leaving behind in the African silence the dead weight of the carcass, the creation of millions of browsing, sun-filled mornings, as the dependent calf emerges from the thicket and stands by dumbly to await the lion.

We head northeast into dry grassy hills. Big pink-lavender grasshoppers rise and sail away on the hot wind, the burring of their flight as dry and scratchy as the long grass and the baked black rock, the hard red lateritic earth, the crust of Africa. To the west rise rough black escarpments, and beyond the escarpments an emptiness in the air, arising from the depression of the Luwegu's valley. Toward the southwest border of the reserve, shrouds of dull smoke ascend to the full fire clouds, all across the Mbarika Mountains.

The path descends into inland valleys of dry thorn scrub and long clinging stands of shrub combretum, then small sand rivers of sweet musky smells and cat-mint stink where crested guinea fowl, a shy forest species, run away cackling under the thickets. But there is no water, and because of their extra loads of meat and the strain caused by the encounter with the rhino, the porters are already tired before noon. To Brian's annoyance, they keep falling far behind. "It's very easy to get lost in bush like this," he mutters. "They must stay together."

Beneath borassus palms a small trickle of good water runs along the bed of a karonga, but under the red banks a juvenile elephant is swatting its legs with a fan of fresh-broken branches, and across the ditch at the wood's edge a cow moves back and forth in agitation; then a second cow with a smaller calf moves into view and disappears again. It is hard to tell how many elephants are here, or where they are, but the nearest among them are much too close. Uneasily, the porters set their loads down, all but the small wide-eyed Shamu, who stares astonished at the elephants like a little boy. Goa says sharply, *"Tua!"* (literally, "to land")—"Put your load down!" And Shamu does so as the second cow emerges partway from the nearest thicket, ears flared out, and drives us back with a loud blare of warning. The confused calf in the karonga tries clumsily to climb the bank, and a third elephant, until now unseen—doubtless the mother of a calf we have not seen either—comes for us across the karonga. *"Kimbiya!"* Goa tells the porters. "Run!" The Ngindo scatter off into the bush, and the rest of us back up rapidly for the second time today. But the cow has stopped behind that wall of vines; we wait and listen while a boubou, startled out of its bush by the elephant, flies across the karonga and resumes its duet without a care.

The calf in the karonga is safely away, and the other calves are, too; we do not see the elephants again. Those in the bush right in front of us have simply vanished, so silently and so magically that even Brian can't quite believe it, and goes poking about in the bush in a gingerly way, just to make sure. Meanwhile, the porters have traveled so fast and so far that we have trouble reassembling them. "We lost track of ourselves," Mata admits, in the wonderful translation of Kazungu.

We cross the karonga, drink warm water, and rest in the cool green shadow of an afzelia. Brian is pensive. "Have to be very careful with these animals, more careful than usual. I don't want to provoke a charge if I can help it. We're a long way from help, you know, if somebody gets hurt, and there's no track for bringing in a Land Rover. These emergencies can happen very fast, even when you see the animals and take pains with them, as we did here." He shakes his head. "What you don't want is to have one elephant cut off from the others. You try to make certain that they're all gone past before you push ahead, but sometimes in dense thicket like this, there's one that's slow or old that you don't see, and then there's trouble."

We have hardly started out, toward three, when we run into more trouble. From a deep thicket comes a profound ominous mutter. Exasperated by the failure of the porters to keep up despite their hard experience this morning, the warden begins a low muttering of his own. "What are these old cows so cross about?" he frets, cautioning the oncoming porters to be quiet. At that moment, the hidden elephant gives a loud and scary blare of warning close at hand—too close for the shot nerves of the Ngindo, who drop their loads and flee without further ado. As Goa and Brian exchange guns, and Brian jams cartridges into the chamber, a monkey gives its loud *yowp* of alarm, and there comes a single sharp loud crack of breaking wood, then a dead silence. After a few minutes, when nothing happens, the porters are whistled for and come in quietly, one by one, grab up their loads, and flee again, loads to their chest.

Our recent adventures with animals have broken down some of the formality between blacks and whites, the separation between their safari and ours. Kazungu tells me that he has never been on a foot safari before, nor have any of the young porters except Mata. Through the big voice of Kazungu I try to explain to the young Ngindo how precious this wild country is, where the water is clean and the game plentiful and even the dread rhinoceros is so peaceful. All the young porters nod fervently, saying *"Ndio, ndio, ndio,"* and Abdallah says, "It is good to see our country, and where the animals are staying," at which they all murmur *"Ndio!"* once again. Kazungu has kept up his journal:

About 9 A.M. we suddenly met a rhino. . . . To avoid being attacked by this rhino, we had to go slowly backward. I was very frightened. For a few minutes, nobody knew which way to go. . . . This was the first time I saw a rhino face-to-face.

After leaving the rhino we went right down to the river to make the porridge, then resumed our safari. The people were behaving well, there was no trouble, they were more attentive now to what they were told, they were polite. . . .

As we were proceeding with our safari, we met some elephants resting. When they realized we were around they started screaming, and we were shocked, and ran away and climbed on the trees, leaving our belongings behind. I knew I would be all right because I stayed close to the person carrying the gun. Everybody ran away except Goa, Bwana Niki, Bwana Peter, and me.

Of the afternoon's encounter, Kazungu wrote:

We could not see this elephant, could not see or hear anything with our ears and everyone was trembling. The elephant started screaming very highly, and this time the belongings were scattered around, and even I dropped my luggage and ran away. Later we started calling to see if we could find each other.

After all this, we went down to the Mbarangandu.

Sand rivers, green river thickets, and dry black cotton marsh, grassy ridges and airy open woods, yellow and copper, red and bronze under blue sky—the *miombo* woods are bursting into multicolored leaf, well before the onset of the rains. . . .

Along the wood edge, a female kudu steps out into bright sunlight. She is crossing one of the black granite platforms inset like monuments in the pale grass, so that even her delicate hooves are clearly visible. A second doe, already off the rock, awaits her. Then a magnificent bull kudu—all adult males of the greater kudu may be called "magnificent"—moves in the same slow dreamlike step over the rock. The kudu are upwind and do not scent us, nor have they heard the sound of our approach. They fade into the woodland. But where the elephant path tends uphill toward the woods, the bull awaits us; this time he raises big pink ears as if to listen to the oriole in the canopy, then stops and turns and gazes at man for the length of a held breath, displaying all the white points of his face—the white muzzle and cheek stripe, the white chevron beneath the eyes, the ivory tips of the great lyrate horns.

It is late afternoon when white sandbars and the blue gleam of the Mbarangandu appear like a mirage in the parched valley, and near sunset when we reach the river plain. An open point set off by borassus palms, with a prospect of open plains and hills and broad bends of the river, and animals in sight in all directions—here was a campsite of the Old Africa that would have been chosen by those men of unknown color who left the stone tools we found on the black granite, more than a thousand centuries ago.

We bathe in the river, and afterward I fix myself a whiskey with fresh river water and sit propped up on my cot, on a low open bluff under borassus, gazing out across the sweep of sunset water to the plain beyond; I feel tired, warm, and easy, and awash with content. Kazungu brings good buffalo stew, and as the stars appear we listen to a leopard just over the river—big deep coughs, well-spaced and strangely violent in a way that the lion roar is not, followed by that rough cadence so like the sound of a ripsaw cutting wood. Perhaps the leopard is disturbed by unfamiliar firelight across the river; or perhaps, like that gentle rhino, like the tame antelopes of the southern Selous, it is only dimly troubled by our intrusion, having had no experience of man.

During the night, hyenas draw near to vent their desolate opinions, and toward daybreak the lions are resounding. "Never heard them at all," the warden grumps, sipping his tea in the gray-pink light of the dawn sky; he has slept badly on the narrow camp cots that in recent years have replaced the sturdy safari cots of other days. "Don't like to miss the lions in the night. Never get sick of that sound, no matter how often I hear it."

In the red sunrise, a pair of pied kingfishers cross the path of light on the shallow river to the palm fronds over-head, and mate in a brief flurry in the sun's rays. With the light, the ground hornbills are still, and doves and thrushes rush to fill the silence.

A cool wind out of the south; we head downstream. A herd of impala, the emblematic antelope of Africa, springs away over the green savanna, and as we pass, the great milky-lidded eagle owl eases out of a thick kigelia and flops softly a short distance to another tree, pursued by the harsh racket of a roller. Already tending north-northwest toward its confluence with the Luwegu, the river unwinds around broad sandbars and rock bends, and wherever it winds away toward the east, the African called Goa cuts across the bends, following the river plains, the hills, the open woods, and descending once again to the westward river.

EPILOGUE

The African elephant, *Loxodonta africana*, has been seriously imperiled by ivory hunters; recent analyses of market tusks show that the poaching gangs, having reduced the savanna or bush elephant, *Loxodonta africana africana*, to less than half a million animals, are increasingly concentrating on the much smaller forest race, *L. africana cyclotis*. Unlike *L. a. africana*, which is easy to census by light plane, *cyclotis* spends most of the daylight hours hidden in the forest, and estimates of its numbers have been mainly speculative. Proponents of the ivory trade maintain that the forest canopy hides very large numbers of small elephants, while ecologists fear that in this inhospitable habitat the numbers have always been low. . . . The future of *Loxodonta* may depend on an accurate estimate of the numbers of the forest race, which would lay the foundation for a strong international conservation effort on behalf of the species as a whole."[1]

In the early winter of 1986, I set off with Dr. David Western on a light-plane journey to the rain forests of central Africa, paying special attention to the numbers and distribution of the forest elephant. From Nairobi, we flew across Uganda to northeastern Zaire, where an air survey was made of the remnant bush elephant and white rhinoceros in Garamba National Park on the Sudan border. In 1961, when I first saw white rhinoceros at Nimule, in the south Sudan, an estimated one thousand to thirteen hundred white rhino lived in Garamba. By 1986 there were fewer than twenty. In an hour's flight over the northern part of the park, we were lucky to find three animals, a lone male and a cow with calf. "The huge, calm, pale gray creatures with their primordial horned heads might have been standing on the plains of Oligocene seventy million years ago, when they first evolved. Except for a lion rolling on its dusty mound, they were the only creatures at Garamba that did not flee at the airplane's approach."

From Garamba we continued westward to Central African Republic and Gabon, where we investigated forest-elephant numbers in the rain forests of the Congo Basin, all the way to the Atlantic Coast. Eventually we returned east across Zaire and spent a week with Mbuti hunters in the Ituri Forest. From there we continued eastward, crossing Lake Edward and Uganda, Rwanda, western Tanzania, and Lake Victoria. On the last leg of the journey we flew low over the Serengeti, from Lake Victoria to the Crater Highlands and Lake Natron. "On a hundred-mile west-east traverse of the whole park, not a single elephant was seen where years before I had seen five hundred in a single herd."

In 1970, as described in *The Tree Where Man Was Born*, the problem in East Africa was too many elephants; since then, 80 percent of them have been destroyed. In 1980, as recounted in *Sand Rivers*, Tanzania's remote and vast Selous Game Reserve held an estimated 110,000 elephants; that number was halved by the time of our forest-elephant survey six years later. Dr. Western's dark conclusions from that survey, presented in 1989 at a CITES conference (the 102-nation conference on international trade in endangered species) in Lausanne, Switzerland, led directly to a worldwide prohibition on the ivory trade. That ban, which became effective in January 1990, is the last great hope of *Loxodonta africana*.

"Of all African animals, the elephant is the most difficult for man to live with, yet its passing—if this must come—seems the most tragic of all. I can watch elephants (and elephants alone) for hours at a time, for sooner or later the elephant will do something very strange such as mow grass with its toenails or draw the tusks from the rotted carcass of another elephant and carry them off into the bush. There is mystery behind that masked gray visage, an ancient life force, delicate and mighty, awesome and enchanted, commanding the silence ordinarily reserved for mountain peaks, great fires, and the sea."[2]

1. All quoted passages in this section are excerpted from *African Silences* (Random House, 1991), unless otherwise indicated.
2. From *The Tree Where Man Was Born* (Dutton, 1972).

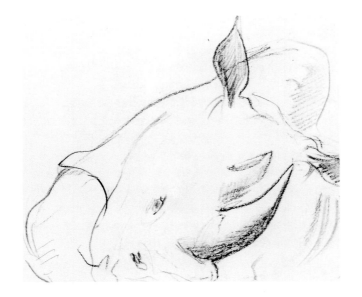

ACKNOWLEDGMENTS   Although Peter Matthiessen and Mary Frank had often talked of collaborating on a book, the idea was not realized until Anne Noonan, then the proprietor of an independent press, approached Matthiessen with a proposal to make a letterpress edition of some aspect of his work. When the project developed to the point that it was no longer practical for letterpress, Abrams became the fortunate recipient of the book in its final form. The authors and publisher gratefully acknowledge Anne Noonan's valuable contribution to the creation of this volume.

PHOTOGRAPH CREDITS   The images in this book, created by Mary Frank between 1970 and 1992, consist primarily of pencil, charcoal, and/or ink-and-wash drawings, with the addition of paper cutouts, monoprints, clay reliefs, and oil paintings on glass and metal. Most of the works are untitled, with the exception of the following: *Expulsion* (1989): 99; *Habitat I* (1988): 102; *Habitat II* (1988): 102; *Migration I* (1987): 56; *Migration II* (1987): 57; *Realm* (1985): 100; *Then and Now* (1985): 58.

The publisher would like to thank the artist for making her art available for reproduction. Additional photograph credits are: Karen Bell: 56–57; John Dolan: 5, 13, 19, 24, 49–55, 58–59, 65, 71, 77, 82, 97–101, 102 above, 104–5, 120; Ralph Gabriner: 99, front and back jacket; Copyright Zindman/Fremont: 1, 2–3, 4, 6, 7, 9, 10, 11, 14, 15, 16, 17, 20, 21, 26, 29, 30, 32, 34, 37, 40, 41, 42, 45, 61, 66, 72, 78, 79, 81, 87, 88, 91, 94, 102 below, 103, 107, 110, 115, 118, endpapers.